NATIONWIDE ANAC
107 32 1404 988%64 633

Improvised Radio Jamming Techniques

Tue 15 August 1995

To George Wallace -
Even fighter pilots need to hear what the "troops" on the ground are talking about. happy hunting...

LAWRENCE MYERS

Lawrence W. Myers

Improvised Radio Jamming Techniques

ELECTRONIC GUERRILLA WARFARE

PALADIN PRESS
BOULDER, COLORADO

Also by Lawrence W. Myers:

Counterbomb: Protecting Yourself against Car, Mail, and Area-Emplaced Bombs

Improvised Radio Detonation Techniques

Smart Bombs: Improvised Sensory Detonation Techniques and Advanced Weapons Systems

SPYCOMM: Covert Communication Techniques of the Underground

Improvised Radio Jamming Techniques:
Electronic Guerrilla Warfare
by Lawrence W. Myers

ISBN 0-87364-520-0
Printed in the United States of America

Published by Paladin Press, a division of
Paladin Enterprises, Inc., P.O. Box 1307,
Boulder, Colorado 80306, USA.
(303) 443-7250

Direct inquiries and/or orders to the above address.

Contents

BACKGROUND

PHASE ONE: TARGET INTERCEPTION

PHASE TWO: TARGET ACQUISITION

PHASE THREE: JAMMING

APPENDICES AND INDEX

BACKGROUND

Introduction

The purpose of this manual is to provide unconventional-warfare (UW) personnel with several expedient methods of defeating enemy communications capabilities in a tactical warfare environment. It is a considerable advantage in any military operation to deny the enemy effective use of the radio-frequency spectrum. Properly timed jamming transmissions that obscure or obliterate radio traffic between operational bases and field personnel can create a significant tactical advantage for a covert, small-unit guerrilla warfare operation.

Large military units employ numerous electronic-warfare elements in the area of countermeasures, whereas small unconventional-warfare teams are limited in this capability by the very nature of their size and operational techniques.

This manual will provide Special Forces personnel with techniques that can be employed in the field to disable enemy communications, using materials and equipment that will generally be available in the area of operation. It will describe how to create a small intercept/analysis/jamming operation targeted toward a specific radio-communications network. The elements of this underground operation are similar in tactics and organizational structure to a large military operation, except that the equipment and manpower requirements are on a smaller scale.

The ideal candidates for this operation on a Special Forces A-Team level would be a communications NCO and an intelligence NCO working together to intercept and identify the target radio traffic. The UW team must coordinate its efforts with other friendly units in the area of operations so as not to interfere with friendly radio traffic on the same bands as the target frequencies.

Radio jamming in the field is a somewhat sophisticated technical problem. It is not one that can be solved by operatives with a minimal understanding of electronic-communications technology. This manual is designed to provide a basic understanding of the principles of radio communications, and to outline methods of detecting, analyzing, and defeating enemy communications traffic for a specific period of time.

Any radio-communications network is inherently vulnerable to electronic countermeasures, or "jamming." The conditions, equipment, frequency band, and built-in protective circuitry of specific radio systems each require consideration and study in order to effectively defeat the entire network during the desired times. It should be stressed that temporarily defeating radio communications can only be done with a specific time frame in mind. Jamming military or police communications is taken very seriously by all governments—the detection and defeat of the jamming device is generally swift and aggressive. Therefore, the planning and timing of a jamming operation is vital to achieve maximum effect on the target network.

Different types of radio communications present different technical problems for the operative. This manual will not attempt to consider all situations or conditions where radio is used; instead, it will focus on the following scenarios:

Law-enforcement communications. This area will be stressed because of its similarity to military short-range communications. UW personnel operating in urban areas have found a significant tactical advantage in having the capability to defeat local police radio traffic for specific durations during the action phase of an operation. It has proven especially useful during hostile extractions and kidnapping missions.

Air-traffic communications. Civilian or military air-traffic communications are especially vulnerable to electronic countermeasures. Disrupting or distorting air-traffic radio can have an immediate impact on the target air terminal. Public awareness of false or deceptive air-traffic radio communications can create havoc at major commercial airlines. Once detected, this type of jamming will result in the immediate suspension of all flights at the target air terminal. This technique can cause panic and economic disruption for an airline, or can shut down an airport for a day.

Broadcast-radio communications. Disrupting a commercial or government radio station is relatively simple. When your team wishes to create hysteria during a small-scale takeover of a target city, jamming the local radio stations can be helpful in convincing the local populace that the government is failing before you actually mount your assault.

This manual is written in a nontechnical style for the urban guerrilla-warfare team. A brief description of basic radio concepts is included to provide the user with an understanding of the inherent vulnerabilities of modern radio communications. Descriptions, reviews, and sources for more than fifty commercially available radio systems are described so the operative (with only a working knowledge of the technical aspects of radio electronics) can quickly rig an intercept station and jamming unit.

A brief historical review of military and terrorist radio-jamming operations is included in order to define some of the applications of this highly useful capability in guerrilla warfare. Recent case histories of criminal jamming incidents in the United States are also included to help assess the potential risks and avoid the mistakes that some groups and individuals have made.

Since modern guerrilla warfare is often staged in an urban setting, the target radio networks discussed in this manual will typically be located in a medium-to-large town in an emerging nation.

However, a study has been conducted on the vulnerabilities of foreign police, military, and government radio-communications systems. These systems bear a striking similarity to those that are set up and operational in the United States. Most of the equipment used in emerging nations has been made or designed in America. The police in most South American countries, for instance, use frequency bands identical to those used by police departments in the United States and Western Europe. Therefore, for training purposes, this manual will discuss the tactical advantages and technical requirements for defeating police, fire, aircraft, and government radio traffic of a typical medium-sized American city.

The focus of this manual is on simplicity, operational security, and expedient execution so that an unconventional-warfare team, using limited resources and personnel, can quickly, covertly, and aggressively attack the opposition's radio communications during an action.

Warning: Possession of this manual can be considered legal grounds for a search of your premises in many jurisdictions. Jamming police radio communications is a considerable risk and potentially dangerous. Commanders are advised to maintain stringent operational security (OPSEC) in the use and distribution of this manual.

Chapter One

Historical Perspective: Radio Jamming in Modern Warfare

In modern warfare, radio jamming is an integral part of every nation's technical arsenal and is classified as *electronic warfare* (EW). Electronic warfare encompasses all elements of radio communications. Jamming operations are termed *electronic countermeasures* (ECM). Military applications for radio jamming are considered effective when the jamming signal affects only the enemy's use of radio and not friendly radio communications. This has not always been the case.

WORLD WAR II

Radio jamming probably began with the first military use of radio communications. Early developments in radio jamming were based primarily on noise-generating devices such as spark-gap transmitters. During the German invasion of Poland in 1939, the Third Reich intelligence group *Abwehr* set out to disable Polish radio communications. Its operation was so effective that German troop commanders temporarily lost radio contact with entire convoys of their own troops during the early phases of the invasion.

All sides used radio jamming during the Second World War. German military strategy included provisions for electronic countermeasures. Japanese forces in the Pacific employed long-range, high-frequency (HF) jamming gear but, like the Germans, they encountered many communications failures with their own units due to the very effectiveness of the jamming signals.

German use of electronic countermeasures became more sophisticated as the war progressed. For example, British Intelligence (MI6) attempted an aerial resupply of Dutch underground units that kept in radio contact with London, called "Operation Northpole." German Abwehr and Gestapo troops were able to capture or kill most of the parachutists from MI6 by locating and capturing the Dutch radio-communications sites and transmitting deceptive instructions to the British teams as they attempted to infiltrate by air. Such radio-deception techniques are termed *black radio*, or *imitative communications deception* (ICD), and it is considered an extremely useful tactic when discreetly employed.

The United States Office of Strategic Services (OSS) both used and fell victim to radio jamming and deception operations. It was the OSS that developed a radio-jamming device that could be deployed by airdrop. This battery-operated unit could be tuned to jam a specific frequency, and it would stay on that frequency regardless of environmental conditions. Its deployment caused significant numbers of "electronic fences" in Axis territory without affecting OSS operatives' ability to communicate with rear-area controllers.

Since the 1940s, the use and deployment of electronic countermeasures have reached a degree of sophistication that gives the user the ability to jam only the desired bands in the radio spectrum, thus allowing friendly radio transmissions to continue unhindered during the interruption of enemy radio traffic.

VIETNAM

Vietnam serves as the most recent example of the effects of radio jamming and ICD operations on the modern battlefield. NVA (North Vietnamese Army) personnel skillfully used U.S. Army radios to call fire-warning orders and bombing requests directly over American positions. Their mastery of American dialects and slang expressions and their knowledge of grid coordinates, authentication codes, and tactical frequencies were based on constant monitoring of U.S. radio traffic by covert listening posts set up and developed by Soviet-sponsored intelligence operations. American casualties from "friendly fire" will always serve as a warning to communications specialists who need to provide field units with secure, reliable communications.

The U.S. Army began to penetrate and defeat these ICD operations during the later years of the war. *Electronic counter countermeasures* (ECCM) were developed to provide a secure and operational radio spectrum during enemy deception or hostile jamming operations. ECCM operations included increasing radio security and operator skill, maintaining detailed reports and intelligence on radio jamming incidents, and locating and destroying enemy jamming sites with special search teams. As a result of these operations, NVA jamming and deception activity decreased significantly in the early 1970s.

On December 21, 1969, soldiers from the 25th Infantry Division located an underground communications complex thirty-five miles north of Saigon being operated by NVA electronic-communications personnel. One enemy soldier was killed and twelve others were captured. All of the prisoners spoke fluent English. According to intelligence documents, the improvised jamming and deception equipment included "modified Sony radios, captured American radio equipment, radios of Chinese Communist manufacture, and homemade sets . . . as well as over 1,400 separate radio transmissions written out in longhand in English." The NVA operatives were able to break U.S. codes and grid coordinates in order to request artillery fire on American positions and help infiltrating NVA units avoid artillery fire.

This was the first of several underground bunkers located in South Vietnam that were apparently used by NVA personnel to disrupt or defeat U.S. fire-support radio communications. Captured equipment was described by military intelligence personnel as "crude" and "homemade" in most cases, but it was highly effective in causing an undetermined number of U.S. casualties due to friendly fire.

SOVIET AND NATO ELECTRONIC WARFARE

The Soviet Union has focused extensively on radio jamming. Soviet military intelligence is controlled by the GRU (Chief Intelligence Directorate of the General Staff). The GRU's Radio Technical Intelligence is assigned the task of providing its action arm, *Spetsnaz*, with intercept and jamming equipment. Team-level Spetsnaz units function as a counterpart to the U.S. Army Special Forces A-Team. Their primary mission encompasses infiltration, espionage, and sabotage, including the disruption of enemy communications systems using a variety of sophisticated jamming devices. The U.S. Defense Intelligence Agency credits Soviet Spetsnaz soldiers with being the most capable electronic-warfare specialists of any foreign power. Indeed, these soldiers may have trained elements of the North Vietnamese Army in radio jamming and communications deception during the war in Southeast Asia.

All current NATO training doctrine regarding radio communications places particular emphasis on potential jamming problems and the means by which they can be defeated. Counterterrorist operations must have excellent electronic-communications units in order to circumvent the extensive radio-jamming techniques being deployed by terrorists.

TERRORIST USE OF RADIO JAMMING

Radio jamming has been used in terrorist operations for several years. The first documented terrorist training program for radio-jamming techniques was discovered by the West German counterterrorist unit GSG9 (Border Protection Unit Nine) when it raided a facility belonging to the Heidelberg Socialist Patients Collective (SPK). The "patients" were under the care of Dr. Wolfgang Huber and his wife, Ursula.

The patients' "therapy" included martial arts training, learning how to build and deploy radio-detonated explosives, and modifying transmission equipment to jam police radio traffic. The patients were also trained in photography and intelligence operations. They photographed police buildings and equipment to learn how to defeat their communications. Finally, Ursula Huber trained the patients in electronic surveillance, and they apparently bugged several government facilities as a part of the training process. This group worked closely with other West German terrorist operations in the use of improvised electronics, so it can be assumed that Baader-Meinhof personnel were technically skilled in radio jamming as well.

Terrorists often use radio jamming as a part of their fund-raising operations. For instance, many South

American and European terrorist groups generate income from bank robbery, kidnapping, and extortion. These activities have a greater success rate when law-enforcement communications are disabled during the action phase of the operation.

Another apparent use of radio jamming by terrorist units in the past several years has been the interruption of emergency medical traffic during bombings and other attacks on civilian or "soft" targets. Upon detonation of an explosive charge in a heavily populated area, terrorist operatives carry out intensive jamming of medical evacuation frequencies to confuse and disable ambulance and emergency medical personnel. This substantially increases the casualty count because it denies quick trauma care to bleeding and burn victims.

This technique is in accordance with one of the major objectives of terrorism—to gain identity and coverage from the press. A high number of victims during a bombing creates more press coverage, and a well-trained jamming team can actually double the number of victims with their disruptive techniques. A secondary effect of this tactic is to cause the press to question the responsiveness of emergency personnel.

The standard method used for jamming medical and law-enforcement frequencies is known as *modulated carrier*. This technique is often construed by the target operator as a *mike-keyed* or *units-doubling* condition caused internally and not by an external jamming condition. This, of course, allows the jamming personnel to utilize this technique without being detected or even suspected of disrupting emergency communications.

COUNTERTERRORIST JAMMING OPERATIONS

Counterterrorist operations have enjoyed considerable

success with their own radio-jamming techniques. For example, the Irish Republican Army (IRA) is perhaps the most adept terrorist group in the use of radio-detonated explosive devices. The casualty rate among civilians and British soldiers is in the hundreds each year. The British Special Air Services (SAS), however, has analyzed fragments of exploded devices and determined the detonation method and even the radio frequencies used. The SAS now deploys several collection and jamming teams throughout its operational area in Northern Ireland. These teams use highly sophisticated spectrum analyzers and receiver equipment to locate radio-controlled bombs and set them off as soon as they are armed by IRA operatives. This has caused such a high casualty rate among the terrorists that IRA technical personnel have been forced to install timers on the receiver sections of their explosives in order to give installation teams adequate time to clear the blast zone before they arm the devices.

IRA bomb makers have since included another device in their explosive packages to detect SAS spectrum-analyzer equipment used to locate bombs. The device seems to detonate the bomb if its circuitry is exposed to nearby scanning equipment. This has caused numerous civilian casualties in Northern Ireland in 1987 and 1988. IRA press releases are now blaming the British military for purposely setting these devices off with their countermeasures equipment, which is certainly technically possible, although the IRA has not made the technical modifications that would prevent the premature explosions. This is perhaps intentionally neglected so the press-hungry IRA can continue to blame the British.

RADIO JAMMING IN THE UNITED STATES

Jamming of law-enforcement and public-service frequencies in the United States has occurred over the past several years. The FBI's VHF (very high frequency) repeater in California has been jammed several times, and the Federal Communications Commission (FCC) has taken part in the investigation to locate the perpetrators. The Orland Park, Illinois, police department was harassed, jammed, and sent on false calls in 1986. The FCC located an 18-year-old ham radio operator and he was criminally prosecuted by the state. The fire department in Pittsburgh, Pennsylvania, was jammed by someone playing tape recordings of Adolf Hitler's speeches on their repeater frequencies. No one was ever caught.

Several northeastern airports have reported false transmissions attempting to misguide commercial aircraft landings in 1988. Again, no one has been caught. In 1986, a Blackhawk helicopter was flying over rural Tennessee on a training mission. An illegal citizens-band (CB) radio attached to a powerful linear amplifier was operating at such high output power that the helicopter's guidance system was disabled and it crashed. The CB station operator was prosecuted.

All of the above incidents depict malicious jamming of critical radio frequencies. It has been speculated that the FCC and other government agencies seldom publicize jamming incidents, and that there are perhaps hundreds of cases that don't get to the news and wire services.

The U.S. government does its own jamming of certain radio traffic. From covert obstruction of foreign embassy transmissions to massive jamming of Radio Moscow and Radio Havana Cuba broadcasts to Western Europe and Central America, these CIA and NSA (National Security Agency) jamming operations are conducted for U.S. national security or foreign policy objectives.

Law enforcement use of radio-deception operations has

become a reality in the 1980s. The FCC has provided several law-enforcement agencies with licenses to operate unattended radar devices on certain stretches of highways and interstate roads. These special devices transmit a strong radar signal toward oncoming traffic, and the popular consumer radar detectors found in many cars alarm drivers and cause them to slow down, which in turn causes other drivers to slow down. Deceptive radar transmissions are an excellent means of creating law-enforcement presence and slowing down drivers without any commitment of manpower, all at a very economical price. Although it may be an exception, this is one case where radio jamming and deception operations are actually saving lives.

PIRATE RADIO OPERATIONS

Another activity that causes jamming conditions is known as *bootlegging* or *pirating*. Criminal enterprises find radio communications to be highly useful for surveillance, transport, and logistics operations. These bootleggers often operate in the marine or business bands, and their radio traffic has been known to disrupt legal communications.

Marijuana growers in Northern California make extensive use of 5-watt VHF (very high frequency) marine-band walkie-talkies. These quiet, lightweight radios are used to protect grower's gardens as well as to advise other growers of aerial surveillance and interdiction operations being sent their way. Interstate truck drivers also have been noted using marine-band frequencies for traffic and radar condition reports. A marine radio goes for about $150 retail, and has a range of three to twenty-five miles, depending on conditions and output.

Pirates and bootleggers also use the CB "free band" frequencies, which are above the legal citizens band and below the 10-meter ham band, for covert traffic. Ham radio equipment in the VHF and UHF (ultra high frequency bands have been used by criminal elements as well. Many of these operators use portable low-power units that are difficult to detect. These units are also programmable, so operating frequencies can be changed regularly for security purposes.

International drug smugglers have been monitored on the high frequency (HF) bands for many years. They use powerful shortwave or ham radio equipment to send covert traffic thousands of miles to anonymous operators in their network who are transporting drugs, weapons, or other contraband on the high seas. (Notorious pirate radio frequencies are noted in Appendix B.)

CRIMINAL RISKS AND PENALTIES OF RADIO JAMMING: RECENT CASE HISTORIES

Poor operational security and sloppy technical discipline when intentionally jamming police and federal law-enforcement agencies have resulted in many prosecutions. What is quite surprising is the minimal criminal penalties meted out in such cases. In 1988, the San Diego police department's radio transmissions were being jammed almost constantly by a radio technician employed by KSON radio. The police used FCC search teams to locate the man, who was arrested and convicted of maliciously jamming law-enforcement and emergency-radio frequencies. He was fined $5,000 and sentenced to six months in jail.

Ironically, this same man was convicted of playing rock music over the FBI's radio frequencies in 1987. He received a $1,000 fine and a suspended three-year prison sentence.

Jamming and ICD operations are frequently perpetrated with no apparent criminal intent, so current laws are focused on the prosecution of the prankster and the radio-electronics experimenter. The type and intent of jamming operations discussed in this manual, however, will certainly result in much more severe penalties if the operators are caught.

By separating the jamming operation from the monitor and action elements, you can avoid showing criminal intent in jamming police frequencies in the United States. However, if your jamming results in injury or loss of life, you may receive penalties of up to twenty years. This, however, is quite unlikely. Cases that have cost the government hundreds of thousands of dollars and risked lives have met with small terms and fines. A good example is a recent imitative communications deception incident involving a resident of Wells Beach, New Hampshire, that was quite sophisticated in its operation.

On January 8, 1988, a commuter airliner heard a distress call over Laconia, New Hampshire, from what was claimed to be a pilot of a downed aircraft. The "pilot" claimed he was injured, and that at least one of his passengers was dead, while others were in serious condition. The Navy ordered a P-3 Orion aircraft that was en route to the Brunswick Naval Air Station to begin searching for the downed plane.

The Navy pilot maintained radio contact with the downed pilot for almost four hours while searching through a heavy blizzard for the aircraft. The ICD operator was very knowledgeable of crash and aircraft jargon, and he even had emergency locator transmitter equipment sending false tones to the P-3. He ultimately caused several hundred volunteers, search-and-rescue units, and dogs to comb almost fifty square miles of wooded area in the heavy snow in search of the aircraft.

When no aircraft was found, and the radio transmissions had stopped, the massive search was called off and the incident was determined to be a hoax. However, a man returning home from work in Wells Beach had called an ambulance after hearing someone in his apartment building crying out that he was "hurt and bleeding." The ambulance driver found a man at home with no apparent injury and stating that he had been asleep. The police visited the man the following day when they learned that someone claiming to be hurt and bleeding over a radio emergency channel had caused an intensive search for a nonexistent airplane crash. They discovered radio-transmission equipment, scanners, frequency logs, and police codes in his apartment. The most incriminating evidence, however, was the fact that his transmission equipment was still tuned to the telling frequencies, and his scanners were still programmed to the search and rescue team's traffic. This and the written materials in his home led to the man's conviction in federal court.

Although he cost the government almost one million dollars in search-and-rescue costs as well as risking the lives of pilots, rescue teams, and volunteer searchers, the individual initially was arrested on misdemeanor charges of creating a false alarm and was released on bail. The U.S. attorney finally charged him with the federal crime of providing false information to a federal agency. He was sentenced to one year in federal prison and two years probation and has been ordered to undergo psychiatric counseling after his release. The case is currently under appeal by the man's attorney for insufficient evidence.

In every other country studied, penalties for jamming typically range from life in prison to execution. Should jamming become more of a problem in the United States, the laws and criminal penalties will probably increase in scope. All of the cases studied for this manual indicate that prosecution was successful only because of the amazingly crude or nonexistent precautions implemented by the operators. It would almost seem that the individuals actually *wanted* to be apprehended and "credited" for their technical skill in causing the incident.

This manual stresses OPSEC and technical precision to avoid detection and capture by the opposition. You should at least understand the current scope of criminal prosecution for these operations. The laws are clear, and the detection and RDF (radio direction finding) capability of the federal government is sophisticated, but the certainty of capture, prosecution, conviction, and confinement is questionable. Current laws do not appear to be very strict or enforceable.

MODERN ELECTRONIC WARFARE CONDUCTED BY THE U.S. ARMY

The United States Army employs electronic-warfare and signal-intelligence teams (SIGINT/EW) at the division level to assist the division commander in his operations. The DC generally maintains operational tasking authority and control over his SIGINT/EW assets. The SIGINT/EW teams are grouped by their operational function. Each division is assigned the following teams (numbers in parentheses refer to the number of teams per division):

1. Voice Collection Teams (6). These small, self-contained, and self-sufficient tactical units are tasked with complete monitoring of the entire radio frequency spectrum. They provide command with audiotape recordings of all identified enemy voice traffic on a real-time basis.

2. *Transcribe/Analysis Teams* (3). These units take the collected voice data and translate it into English, transcribe the data to text, and then provide a nontechnical description and a technical analysis of the content of the traffic.

3. *Communications Jamming Teams* (6). These units are assigned specific frequencies and bands to jam. They are given priority frequencies for continuous jamming, as well as timed jamming on specific target frequencies for specific durations.

4. *Noncommunications Collection Teams* (3). These units focus on radio traffic that is used for guidance and radar-controlled devices, such as counterbattery radar, ground-surveillance radar, target-acquisition radar, and special weapons. These teams provide telemetry data to the Transcribe/Analysis teams regarding the purpose and effectiveness of enemy weapons control and radar systems.

5. *Multichannel Communications Collection Team* (1). This unit is assigned the task of intercepting pulse-coded modulation (PCM) traffic from enemy signal centers. PCM transmissions contain several dozen conversations or data signals on one modulated carrier; thus they are termed multichannel communications.

The collection, analysis, and jamming teams are supported by a flight platoon of three helicopter units that provide rapid mobility and increased range. These four operational elements comprise the typical EW company assigned to a division commander.

Radio interception techniques are used quite often in today's unconventional warfare environments. A good example of this is the current operation being run by U.S. Army Special Forces personnel in El Salvador. They have set up a series of "tracking stations" in the mountains equipped with scanners and directional antennas. Using RDF triangulation techniques (see Chapter Eight), SF radio technicians are able to locate leftist guerrillas operating hand-held VHF walkie-talkies. By estimating that there is one radio for every eight to ten guerrillas, an approximate enemy strength count can be determined.

A growing trend in electronic warfare is the use of imitative communications deception methods, particularly in the low-intensity conflicts of the last two decades. ICD operations are extremely dangerous, primarily because the ICD personnel generally must be close enough to enemy positions to duplicate their traffic. The success of ICD operations depends on knowledge of the target's communications, compatibility of available equipment, and the language qualifications of the ICD operatives. U.S. Special Forces personnel have had excellent opportunities to demonstrate their stealth and language skills in ICD operations around the world.

Chapter Two

Basic Principles of Radio Jamming

There are three basic methods of radio jamming: spot jamming, barrage jamming, and sweep-through jamming. The jamming signals are referred to and classified as modulated or continuous wave. The jamming of radio traffic is termed *electronic countermeasures* (ECM) in tactical communications terminology.

Any transmitter can be made to function as a radio-jamming device. If the transmitter can operate on the target frequency, it can be used as a spot-jamming device by transmitting one narrow-band signal directly over the enemy's traffic.

All electronic systems that depend on reception of electromagnetic radiations are vulnerable to jamming. This includes radio communications systems (voice or data, continuous wave such as Morse code, and radio teletype) and radar systems (such as proximity fuzing missile payload systems, missile and aircraft guidance systems, navigational-aid systems, and telemetry systems).

The UW team must realize that the use of portable FM

or UHF transceivers by military and police patrols is different from the technical applications of radio equipment used for air-traffic control, forward observer artillery or fire control, and in electronic guidance systems. In order to effectively disable a target communications network, the UW team must first identify the type of communications equipment deployed by the enemy and the frequencies being used. Once this is done, monitoring their traffic for a period of time will provide the UW commander with an understanding of the enemy's particular use of radio in its daily operations and overall mission.

It is vital that the UW commander understand that all military and law-enforcement operations constantly depend on radio. Yet these organizations can and do function effectively with temporary losses of radio communications. Most organizations employ *radio silence*, or loss of radio-communications capability, in their training doctrine in order to reduce a unit's vulnerability to countermeasures being employed against them.

The purpose of jamming is to create and sustain confusion among patrol and field elements, not to disable the entire organization's effectiveness to the extent that it is completely unable to respond to the UW teams' action. Jamming is best employed as a means to limit a target unit's response to an action or to stop backup units from being aware of the need for their deployment during critical phases of the UW team's planned action.

RADIO PROPAGATION

Radio propagation is the behavior and effect radio signals have in the atmosphere. Most radio transmissions that the UW operative will be attempting to defeat will be line-of-sight communications. These transmissions generally are limited geographically because the signal must take a direct path to the receiver.

Radio waves are sent from the atmosphere through an antenna. The physical size and characteristics of the antenna are determined by the desired operating frequency, the range and power of the transmitter, and the geographical conditions of the radio site.

Certain radio waves can be sent over long distances—in fact, completely around the world—due to the condition of the Earth's atmosphere. A shortwave radio receiver can pick up broadcasts from anywhere in the world because the signal actually bounces off the ionosphere and returns to Earth several times, enabling it to be heard anywhere on the planet. Radio transmissions in the 2.5 to 30 MHz (megahertz) range can use the ionosphere to send and receive traffic over these great distances. Special Forces radio equipment operates in the HF band and uses sky-wave propagation to send encoded traffic over long distances.

Figure 1 illustrates the effects of certain radio signals in the atmosphere. Weather, time of day, and time of year all have an effect on radio propagation as well as on the operating frequency of the radio traffic.

RADIO TRANSMITTERS

Electrical energy is measured in *volts* and *amperes.* Voltage is the measurement of the force of electric current. The higher the voltage, the higher the force of the electricity. Current is the actual movement of electrons through a conductor. Current is measured in amperes.

There are two basic types of current: *alternating current* (AC) and *direct current* (DC). Alternating current constantly changes its voltage from positive to negative. AC is what comes out of your wall socket. The speed of the alternation of the AC is known as its *frequency,* and is

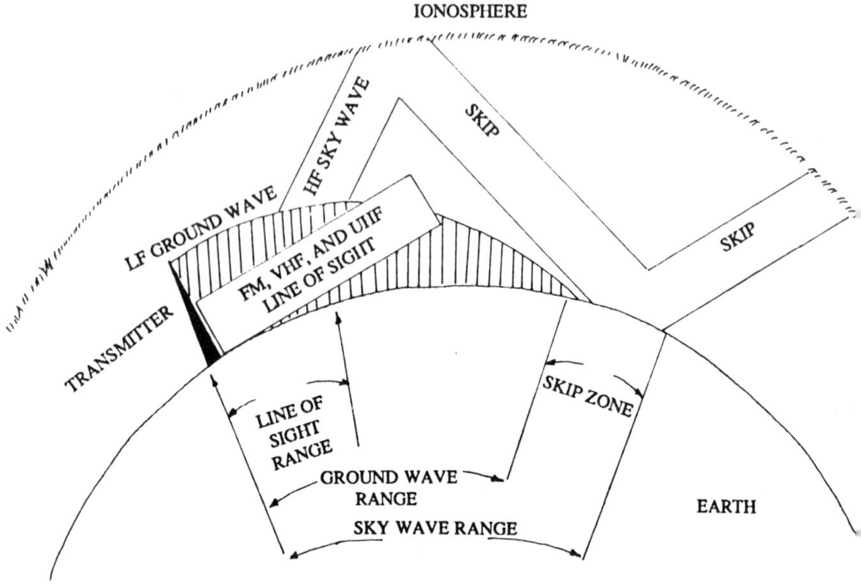

Figure 1. Radio signal propagation. Low frequencies (150 kHz-400 kHz) are considered to be ground wave and travel several hundred miles. High frequencies (2MHz-30MHz) are considered to be sky wave and have an unlimited range. These signals repeatedly bounce off of the ionosphere and back to earth. FM, VHF, and UHF frequencies (30MHz and above) are considered line-of-sight and normally travel 35 to 50 miles, and up to 200 miles under ideal conditions.

measured in *hertz*, or cycles per second. When electrical current travels through a conductor, it produces an electromagnetic radiation field around the exterior of the wire. If you connect a battery to a light bulb and put a compass near the wire, you can see the needle move when the light is energized. The compass is detecting and responding to the electromagnetic radiation coming from the wire conductors in the circuit.

The speed or frequency of electromagnetic radiation is the means by which a radio receiver is tuned. The electronic circuitry that produces the current at a specific frequency is known as an *oscillator*. An oscillator creates electromagnetic radiation at specified radio frequencies and is the main circuit inside a radio transmitter. The oscillator transmits a signal through an antenna, and the signal radiates from the antenna into the atmosphere.

One form of radio transmission used extensively in military, covert, and amateur communications is known as *continuous wave*, or CW. CW is simply the interruption of the oscillator's output with a keyer, which sends a series of tones in the form of dots and dashes called *Morse code.* Morse code can be sent manually or electronically. It is considered to be the most reliable form of radio communications because it can be copied through static, atmospheric, and weather conditions that prohibit other types of communications, and it can be sent over long distances compared to voice or data from the same transmitter. CW transmitters are also the simplest to construct and operate.

The signal from an oscillator generally is amplified before it reaches the antenna. The current produced from an oscillator is usually very low, and an amplifier is used to strengthen the signal in order to increase the range and the power of the transmission before it reaches the antenna. Since the frequency of the signal is in the radio-frequency (RF) range, the amplifier is known as an *RF amplifier*.

The antenna functions as a radiator for the signal, sending it out in all directions. Since radio signals are electromagnetic in nature, they each have a specific physical size, or *wavelength.* Wavelength is measured in *meters*. A transmitter's wavelength is inversely proportionate to its transmitted frequency. For instance, CB radio signals are in the 27 MHz frequency range and have a wavelength of 11 meters. Some police radio-communication frequencies are in the 150 MHz range and have a wavelength of about 2 meters.

By observing the physical size, position, and type of antenna used by a radio transmitter, the UW operative can easily determine its approximate operating frequency. This technique will be covered in detail in Chapter Eight. At this point, it is important to remember that the antenna is the component in a radio used to radiate and receive radio signals.

Although continuous-wave traffic requires only an oscillator, an amplifier, and an antenna, a fourth component is needed in the transmitter in order to send speech over the airwaves. The process of putting speech on the continuous wave of an oscillator is known as *modulation.* The operator's voice goes into a microphone, which converts the speech into electricity of varied speed or frequency. The frequency of the human voice varies from about 100 to 3,000 hertz in the *audio frequency range.* Audio frequencies are those frequencies that can be detected by the human ear, generally considered to be from 20 to 20,000 hertz. The voice's audio frequency is amplified with an audio amplifier similar to the type used in a stereo receiver. This amplified audio is then combined with the continuous-wave carrier signal coming from the oscillator, usually in the power-amplifier section of the transmitter.

Now the amplitude of the CW carrier signal varies with the changes in speech coming from the modulator. This process is known as *amplitude modulation*, or AM. When the modulator is made to vary only the frequency and not the amplitude, it is known as *frequency modulation*, or FM (see Figure 2). Figure 3 is a block diagram of a simple transmitter, showing the placement of each component in a radio transmitter.

Radio jamming does not actually disable the transmitter electronically or physically. Rather, it disables the transmitter's ability to effectively be heard by units with receivers by transmitting a more powerful signal over the transmitter's signal. Radio jamming is like preventing two people from talking by making a lot of noise so they can not hear each other. It does not actually prevent them from talking.

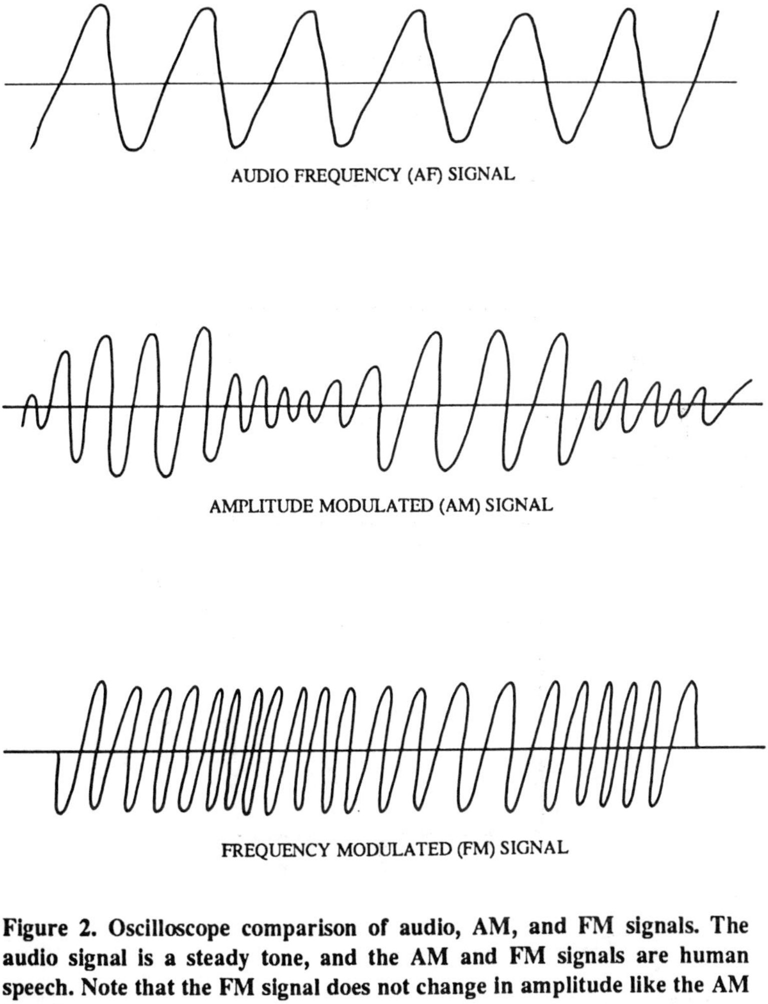

Figure 2. Oscilloscope comparison of audio, AM, and FM signals. The audio signal is a steady tone, and the AM and FM signals are human speech. Note that the FM signal does not change in amplitude like the AM signal.

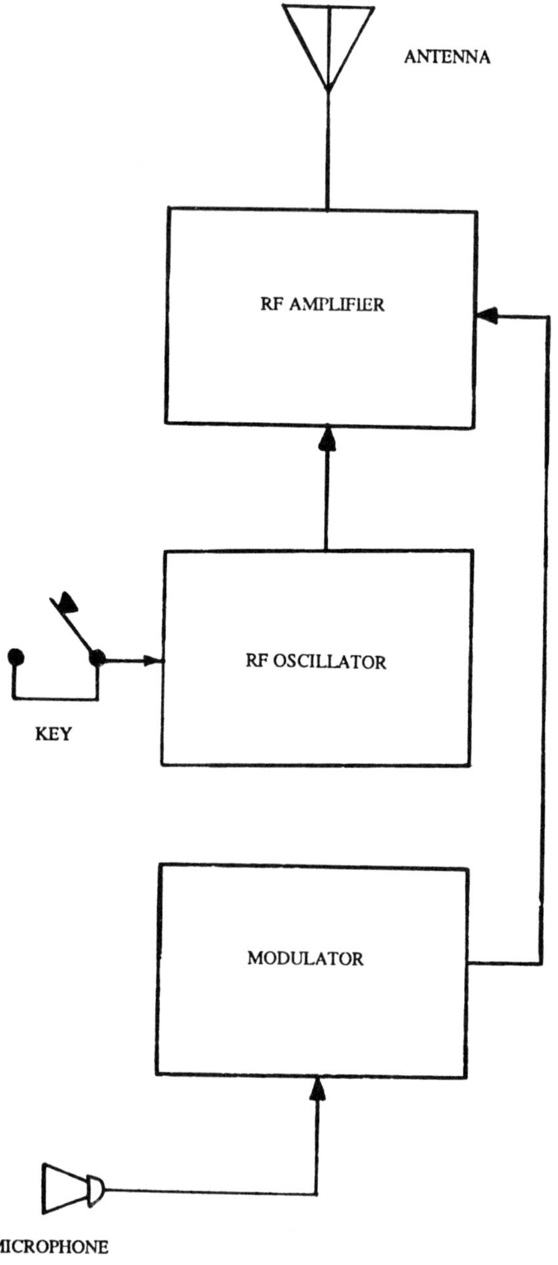

Figure 3. A simple transmitter.

RADIO RECEIVERS

A radio receiver functions almost in the reverse of a radio transmitter. It takes a signal from the airwaves, removes the RF, amplifies the audio signal, and sends it to a speaker or pair of headphones. There are several important components in every modern radio receiver.

The antenna gathers radio signals from the air. As in a transmitter, the size of the antenna is related to the wavelength of the frequency that will be received. Most antennas are not a full wavelength in size, which would not be practical for high-frequency transmissions. For example, a CB radio antenna tuned to the wavelength of 11 meters would be almost 40 feet long. Instead, most antennas measure one-half or one-quarter the wavelength of the desired frequency.

Regardless of the size of a radio receiver's antenna, it actually will pick up thousands of different radio transmissions at one time. Therefore a receiver has a *tuning dial*, which is a circuit that tunes in a specific frequency and avoids all other frequencies. Inside the receiver is a variable capacitor connected to the shaft that holds the tuning knob. Without getting into a technical explanation of this function, suffice to say that the variable capacitor, in circuit with an inductor (which is a tuned coil), resonates at a specific frequency determined by the position of the tuning knob and the variable capacitor.

All modern receivers also have an oscillator circuit, which functions as a miniature transmitter. It performs a vital function for the receiver by making it more sensitive and much more selective, and is known as *superheterodyne radio reception circuitry*.

The radio signal enters the receiver through the antenna and is amplified. It is then mixed with the oscillator's signal, which is known as the *intermediate frequency* or IF signal. The received signal and the IF signal are determined by the tuning capacitor and then amplified further. The signal is then sent to a detector stage, which removes the IF signal, separates the RF from the audio signal, and then amplifies the audio signal so that it can be heard in the receiver's speaker. Figure 4 is a block diagram of a typical radio receiver.

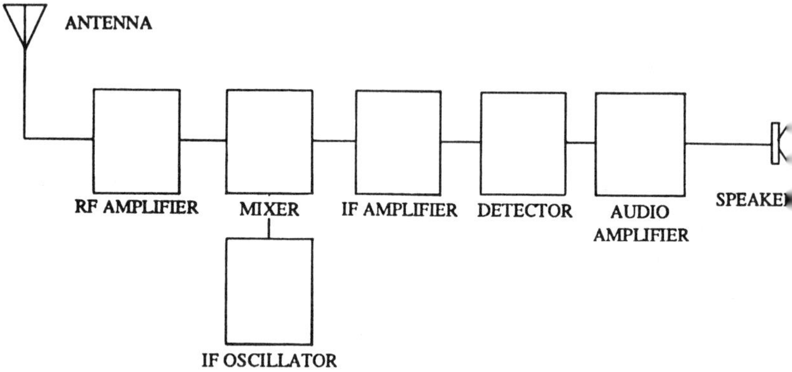

Figure 4. A superheterodyne AM radio receiver.

Again, this is a simplistic explanation of a receiver's function, but it is important for the operative to understand that a receiver is also a tiny transmitter due to the oscillator in the IF section. This IF oscillator is easy to detect using modern spectrum-analyzer equipment—it is what counter-terrorist forces search for when they sweep an area for radio-detonation devices or hidden radio-monitoring sites.

UNDERSTANDING RADIO OSCILLATIONS

It is important to understand the effect the receiver oscillator has on other receivers. As we have noted, the radio receiver also acts as a transmitter. The transmission frequency of the oscillator in a receiver's IF section is fairly standardized. If the radio is a typical AM type, the oscillator will transmit at 455 kHz. Citizens band, shortwave, and standard AM broadcast radios generally have an IF of 455 kHz. FM radios, Walkman portable stereo receivers, walkie-talkies, and VHF scanners all generally have an IF frequency of 10.7 MHz, 10.8 MHz, or 10.85 MHz. (Other electronic devices also produce oscillation. Portable typewriters, personal computers, hand-held calculators, and even quartz digital watches all produce oscillations.)

To see how oscillations from two receivers have an effect on one another, put two AM radios side by side and turn them on. Tune one radio to the middle of the band and begin tuning the other radio toward either end of the band. As soon as the radios are 455 kHz apart in frequency, you will hear a loud whistle or tone coming from each. This condition is known as *heterodyning* and can actually be used to troubleshoot a radio that has a suspected failure in its IF section. The tone indicates a functioning IF section in each receiver.

Anyone who has been aboard a commercial aircraft and turned on a portable radio has probably already learned about IF oscillations and interference between two receivers. It is against FAA rules to operate any radio receiver in the passenger compartment of an aircraft. The oscillations from the receiver can cause havoc in the air-to-ground radio equipment as well as in the navigational and guidance equipment in the aircraft's cockpit.

Most military-grade electronics equipment is shielded in metal enclosures to protect them and other nearby devices from each other's oscillation interference. Shielding digital and computer equipment is also necessary for security reasons. If a computer is improperly shielded, all information typed in, read from memory storage, or printed out can easily be read by a nearby enemy group operating receiver equipment that can pick up the oscillations from the computer's clock and CPU as well as the IF section of the video monitor.

There have been cases where criminals have used this technology to gain access to Automated Teller Machines (ATMs). By monitoring the ATM while it is being used by a bank customer, criminals can learn the customer's access code. The criminal then either steals the customer's card or makes a card of his own. One well-documented case in New York City involved a Polish immigrant who was adept in electronics. He learned several thousand access codes using receiver equipment and built his own ATM bank-card manufacturing machine to take advantage of the knowledge.

As illustrated in the above example, understanding the oscillation functions in radio receivers and most digital electronic devices can provide the electronic guerrilla with a very sophisticated edge in defeating enemy communications. The concept of IF oscillations is a key to the vulnerability of modern electronic devices and the means by which they can be defeated.

PHASE ONE: TARGET INTERCEPTION

Chapter Three

The Covert Listening Post

In order to effectively use radio jamming in a guerrilla-warfare environment, it is essential that the UW team understand the target network and its tactical use of radio communications. In other words, it is important to know how your target uses radio and what he will be unable to do without it. In addition, it is vital to understand how your target would respond when he finds his system being jammed. Therefore, it is tactically and technically advantageous for your team to monitor the target net for a period of time before jamming its radio communications.

This section will discuss how to set up and operate a covert listening post (LP) for intercepting the target radio traffic using modified consumer products and other generally available equipment. The equipment discussed will provide the user with the ability to continuously scan and monitor the entire usable radio-frequency spectrum.

The listening post can be anything from a "hole in the wall" room for rent to an extra room in the team quarters. It should be set up to operate under its own power and in unusual weather or climatic conditions. Monitoring equipment should be battery-operated and portable, making it easy to carry in a pocket or small briefcase that will not draw suspicion from the local population.

Ideally, the covert site should be in a clean, secure environment as physically close to the target as possible, and close to the planned action. The selection of the site will be most critical during the action phase of the operation, when both good signal reception and tight security will determine the mission's success. Listening post site selection is based on the following criteria:

1. Proximity to target, for clear signal reception.

2. Elevation (i.e., high-rise apartment), for better line-of-sight signal reception.

3. Security. Radio gear, with blinking lights, readouts, etc., is fascinating to most people. If a case of radio jamming is publicized in the news, they may remember what they saw.

4. Available power/life support for operators.

It is vital to compartmentalize the overall operation as much as possible. All records of successful jamming operations bear this premise out. NVA operatives in Southeast Asia always kept their monitoring, jamming, and action teams separate and isolated. Terrorist teams in Europe use monitor sites that are in no way affiliated with the rest of the network other than providing them with intercepted voice traffic from their targets. These radio monitors often have no idea what the real purpose of their work is.

Perhaps the most significant advantage of having an isolated listening post is so it can monitor the target's response during the action phase of your team's mission. When the target is jammed, the radio operator will attempt to communicate in other ways. He may have alternate radio frequencies in use, or he may use other communication assets to advise all elements to suspend or reduce their patrols or activities. If the monitor site is separate from the rest of the UW team, it can pick up this traffic to identify these emergency (and perhaps covert) frequencies. Additionally, an isolated LP can monitor the radio spectrum during jamming "test runs" to further study the target net before the planned action is attempted.

This highlights an important point. Once the target or targets have been determined, it is vital to identify *all* potential operating frequencies. Aside from published sources noted in Appendix A and the standard allocations of frequencies listed in Appendix B, there are several other means of determining your target's operating frequencies that will be discussed in the next section.

LISTENING POST OPERATIONS AND PERSONNEL

A well-run monitor station will provide your operation with a big advantage over your target. Its basic functions are:

1. Continuous search of radio spectrum for enemy activity.
2. Tactical voice intercept of all known enemy radio frequencies.
3. Operational planning input of enemy activities and movements.
4. Log all known operating frequencies for jamming purposes.
5. Monitor enemy activity during the planned action.

The more dedicated and attentive your monitor-station personnel, the more tactical benefits they will achieve. Current military intercept operations usually involve two or more personnel. One operator functions as the voice interceptor while the other functions as the search/scan operator. The most experienced voice interceptor is employed as the search/scan operator because he can recognize enemy radio traffic and distinguish it from the other various transmissions that will be encountered during a complete search of the radio spectrum.

The listening post should be manned constantly. It therefore must be comfortable and provide adequate life-support facilities for an intercept team composed of several members to provide round-the-clock monitoring and interception. If this is not possible, then there are options such as sound-activated recording devices and computer radio interface systems (CRIS) that can be employed to allow your station to be unattended at least part of the time.

No operational personnel or secondary operatives should ever be seen near the listening post. The LP operators might be involved with technical support for jamming ops along with their intercept duties, but it must be carefully understood that the site for jamming and the site for monitoring have to be kept completely separated both in location and personnel.

LISTENING POST LAYOUT

The physical layout of a monitor post is very important for a smooth-running operation. All equipment should be placed on a table large enough to allow the operator to make notations and refer to maps or charts during monitoring. The layout should provide access to any individual switch, dial, or button. Earphones and tape-recording equipment should be accessible as well. Some covert monitor sites are located behind a false wall in a closet. This is relatively secure, but it can be very uncomfortable during continuous operation.

Make room under the table for cables, power strips, and your backup battery system. A backup lighting system and a fire extinguisher are also recommended.

A personal computer creates a tremendous amount of electronic noise when it is operating. Therefore it should be kept on a separate table as far as possible from the radio desk. The power cables for the computer should be separate from the receiver power cables as well. Ideally, you should use a separate wall plug in another room for the lowest noise ratio. Finally, certain receivers may conflict with each other, so a degree of experimentation is required for optimum placement of each unit.

The monitor site should be set up for mobility, security, and ease of use. Your intercept team will practically live at this location for days and weeks at a time. The system described in this manual can easily fit into the back of a covered pickup truck or small van. This would provide the extra OPSEC capability of constant mobility.

Chapter Four

Intercept Equipment Selection

A well-prepared listening post is set up to intercept and record many different types of radio traffic for study. Certain types of government radio traffic can appear anywhere in the radio spectrum. Your target agency may use HF, VHF, UHF, or microwave radio traffic, all of which can originate elsewhere and be beamed to your target via satellite. In order to ensure that you have the capacity to intercept all possible enemy traffic, you should be equipped with several receivers.

The advent of the microprocessor has had a significant impact on the sophistication and cost of high-speed scanning digital radio receivers that can continuously monitor hundreds or even thousands of frequencies up and down the radio spectrum. There are many features and options available, as well as simple modifications and enhancements that can be added by the user.

This chapter will discuss digitally programmable high-speed scanning communications receivers that are currently available in the United States and Western

Europe. The cost of a complete monitoring station—with internal and external battery backup, covert antenna system and accessories, and two-way communications for multistation link—can run from $2,000 to $10,000, depending on equipment needs and budget. This section will discuss several systems, with emphasis on low cost, high technology, portability, and ease of operation.

Note: Many of the equipment, modifications, and techniques described in this section are illegal. Most countries, including the United States, place restrictions on the interception and monitoring of certain types of radio traffic as well as certain frequencies. The United States has recently passed legislation that makes monitoring certain types of radio traffic a criminal offense. The Electronic Communications Privacy Act of 1986 (ECPA) makes much of what we will be discussing in this section a crime. In Europe and South America, governments are even more restrictive on this type of activity. This manual stresses the need for careful consideration of the risks involved with monitoring radio traffic.

* * * * * * *

There are several components to an efficient and combat-effective tactical listening post. Once you have selected a secure, covert location and competent personnel, you must then focus on the following operational criteria:

1. *Power Requirements*. Your station should have excellent main and backup power. The wiring should be carefully planned and documented for operator changes. A *commcenter power diagram* should be drawn up and kept on site for reference.

2. *Receivers*. The site should contain several receivers capable of digitally searching through entire bands or

ections of bands for traffic activity. The site should also ontain several units that have high-capacity memories to ontinuously scan all known enemy radio-traffic channels, ither published or as a result of your searching activities. Your station should have tape-recording and tape duplica- ion capabilities. A small computer for frequency logging is lso useful.

3. *Interstation Communication Link.* The site should ave a covert communication link between the intercept perators and command, as well as with specific team lements such as site security.

4. *Covert Antenna System.* Obviously you do not want he roof of your monitor site to look like a foreign embassy, vith several strange antennas mounted on masts. The *ommcenter antenna diagram* should show antenna loca- ions, elements, interconnections, amplifier wiring, etc. Each receiver, as well as your interstation communication ink, will need low-profile dedicated antenna systems.

5. *Logging System.* This can be as simple as a spiral otebook or as complicated as a small personal computer ystem, depending on one's resources and operational equirements. All intercepted radio traffic will be useful if roperly logged and analyzed. Even periods of *no* radio raffic should be logged for future study. Lack of radio raffic can mean many things: an action requiring radio ilence, the enemy's use of alternative means of com- nunication, a system failure, or possibly maintenance lowntime.

6. *System Layout.* The majority of your monitor equip- nent will sit on one or possibly two tables. The location nd position of each device and control should be consid- red for maximum operator efficiency and accessibility to ny one control quickly.

7. *Documents Library.* There are numerous publica-

tions regarding frequencies, agencies, codes, and technica data that will be highly useful in your site operation Monitoring is a growing hobby, and there are magazines books, and manuals that focus on this activity. There are even associations and clubs devoted to monitoring. See Appendix A for a list of sources on this growing hobby.

POWER REQUIREMENTS

A mobile communications site requires battery power or at least battery backup. Planning is required to determine your team's needs in this area. After selecting your equipment and determining factors such as operating conditions, operating voltage, and current consumption, you should draw a power routing and layout sheet. This doesn't need to be elaborate. Just sketch out all of your connections and lines going between outlets and your communications table. Estimate your needs and plan for expansion. Figure 5 is a drawing of the power cable and equipment hookups for a five-receiver intercept station, with two-way radio link, lighting and auxiliary power, and a three-level battery backup system consisting of a 12-volt marine battery, gel cells, and nickel-cadmium (ni-cad) battery arrays.

Basic mobile tactical listening post requirements will include several pieces of equipment, many of which will require different voltages to operate. This will cause a potential problem unless planned into the system. Let's assume that you will require a system capable of operating on 110/220 volts AC conventional wall power, 12 volts DC automotive or backup power, and on internal rechargeable battery packs for individual or further backup capacity.

Most of the equipment used for tactical voice intercept is very low-current consumption equipment. In fact, many pieces of receiver gear will have indicator lights that consume as much or more power than the rest of the

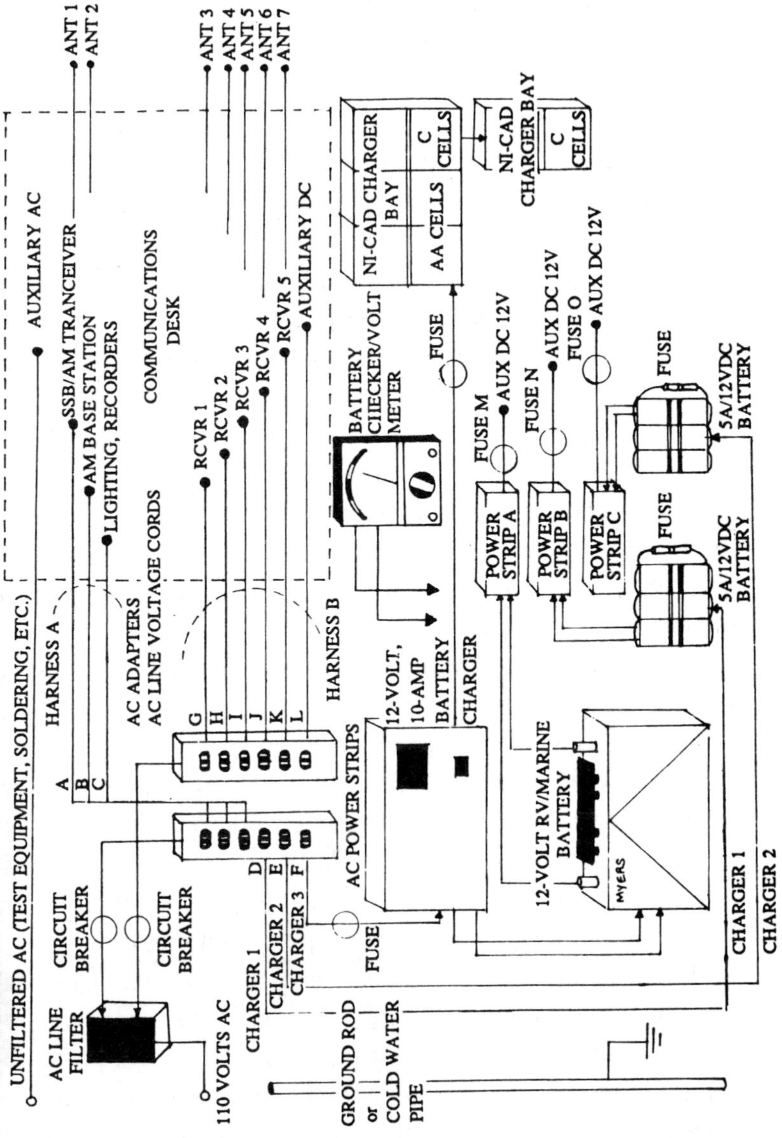

Figure 5. Intercept station power diagram.

receiver. These indicators are not always necessary for efficient or continuous operation. We will discuss modifications that can provide the user with switchable indicator lamps, which lower current consumption and provide better security by improving light discipline.

The first consideration in operating a continuous radio-intercept station is sufficient battery power. For practical purposes, the industry standard for most communications equipment is 12 volts DC. Since most of the receiver equipment is low current, it is best to use a battery that can handle a trickle of current drain for an extended period. A standard deep-discharge marine battery has proven to be the most suitable for this application. These batteries operate trolling motors and radios on boats and are even capable of high-current "cranking power" to start engines and power high-current VHF radio equipment. This type of battery is available at Sears and most discount auto and department stores.

The monitoring station described in this manual uses a Sears DieHard deep-discharge marine/RV battery and a 10-amp charger. This battery is bulky and a potential fire hazard in a poorly ventilated area. It will, however, power the monitor station for almost three weeks before falling below 40-percent charge. Most of the equipment described here also contains space for internal batteries—rechargeable nickel cadmium (ni-cad) batteries are recommended.

Since there may be times when the internal batteries and the main deep-discharge battery will need to be recharged, the monitor site should also include two portable 5-amp hour 12-volt batteries (available at Radio Shack, model #23-182). These units weigh about six to eight pounds and are used to power video cameras and portable electronic equipment. They generally are packaged in a durable nylon case and come with wiring for recharging either from a wall outlet or a 12-volt car battery.

There are special considerations for AC power. First, your monitor site will have eight to twelve different pieces of equipment, all requiring an individual plug. A multiplug power strip is helpful here. There is also the sporadic noise that is encountered in most AC power lines. You will need to obtain an AC line-interference filter (Radio Shack model #15-1111). All AC lines should have grounded three-prong plugs for equipment protection and safety. (*All* of your power capability should be fused or have circuit breakers in line to protect it from fire and equipment overload.)

Other options to consider for power and/or backup are low-cost portable generators and solar panels. Photovoltaic panels can continuously recharge your main battery while another panel operates your monitor station during sunny days. These rather sophisticated power options are expensive and are only practical if you are going to be in the most remote of environments.

The following is an inventory of a practical, efficient power system for a multichannel monitoring station that has good backup and long-term power capability for mobile or tactical purposes (prices applicable at time of publication):

ITEM	SOURCE	COST
AC power strip (2)	any department store	$21.90
DieHard battery	Sears	$65.95
Deep-discharge RV/Marine battery charger	Sears	$39.95

ITEM	SOURCE	COST
10/2 amp interference filter	Radio Shack	$7.95
12V/5 amp battery (2)	Radio Shack	$119.90
TOTAL COST		$255.65

Other costs will include wiring and plugs, jacks, fuse blocks, and connectors. These items usually are supplied with each receiver, but are available as options on others. One very important accessory for the battery is a plastic case. The deep-discharge marine batteries available from most stores come with an optional case. These PVC or ABS plastic cases protect the battery from corrosive elements as well as from stray sparks or cigarette ashes that could cause a fire or explosion. They cost from $8 to $12 and are a worthwhile investment. (*Note*: When recharging these types of batteries, always remove them from the case and charge them in a well-ventilated area away from any spark, flame, or heat. These batteries can explode very easily when they are in the recharge mode.)

Nickel-cadmium batteries are expensive, but they can usually be recharged anywhere from 300 to 500 times, making them much more cost effective than nonrechargeable alkaline batteries. Budget about $100 for ni-cads and about $300 for your main power, bringing your total cost for power and backup to around $400.

RECEIVER EQUIPMENT

Figure 6. Covert listening-post equipment.

The computer chip has revolutionized communications-intercept equipment. This small silicon microprocessor provides us with programmable scanners and receivers, interface devices, and other computer equipment that is now affordable for many organizations.

Military and government agencies assigned the task of radio-voice intercept use special equipment with features and scan speeds that are generally unavailable to the average consumer. For instance, the National Security Agency (NSA) uses some of the most advanced communications-intercept equipment in the world, most of which is classified. The growing trend in intercept technology is CRIS (Computer Receiver Interface System), which allows a computer to control and log several thousand frequencies with little assistance from an operator. The NSA also has computers that can actually identify the voice of the radio user and pick out key words from the traffic. All of this is done electronically, without having the site continuously manned by intercept personnel.

One of the most sophisticated scanning devices used by

military intercept units is the R-2412/U. This scanner is capable of covering 20 to 1,200 MHz at a rate of 50 channels per second. Its modules each have 100-channel memories, and the entire system can literally scan every possible radio frequency known every second. The R-2412/U is manufactured by the Cubic Corporation in the United States. It is obviously very expensive and may not be available to some parties.

If your operational budget is high, Appendix C lists some firms that offer CRIS equipment and other advanced intercept gear. Most guerrilla-warfare units are severely limited in resources, however. For that reason, this section will focus on low cost as well as sophistication, avoiding experimental or overly advanced equipment.

* * * * * * *

Selecting receiver equipment for your monitor site will be based on technical requirements, resources, and equipment availability. Most of the usable radio spectrum will need to be covered. In order to do so efficiently, you will need to have digitally programmable scanning equipment.

A scanner is a multiband portable radio receiver programmed to specific radio frequencies. In order to monitor law-enforcement, military, or government traffic, a scanner allows you to scan through anywhere from 10 to 400 radio frequencies in a couple of seconds. When the scanner receives a signal, it stops on that frequency for the signal to be heard and then continues through its programmed memory of frequencies in search of another signal.

There are many different scanners and receivers, with hundreds of features and options available. There are, however, several desirable features that should be considered for a UW operation:

1. *Frequency coverage.* Your specific targeting needs will determine the necessary frequencies to be covered (Appendix B breaks down the radio spectrum by agency, frequency, etc., for this purpose). Since many agencies operate at least partially in a covert manner, it is vital that the scanning equipment cover as much of the radio spectrum as possible.

2. *Sensitivity.* Scanners are unusually sensitive receivers. They can pick up anything from low-power transmissions from walkie-talkies to distant aircraft transmissions with relative ease. A receiver's sensitivity rating indicates the lowest input signal voltage that the unit will respond to and be able to copy. The lower the sensitivity rating on a scanner, the better. (Some manufacturers do not offer this rating, usually because they are selling equipment that is not very sensitive.) Most of your intercept activities will be undertaken somewhat close to your target. If your receiver is too sensitive when used with one or more antenna systems, you will find the signal desensitized and actually sounding worse than if it were picked up without an external antenna.

3. *Images.* All scanners generate IF oscillations. As we discussed with receiver specifications, these oscillations have an unusual effect on certain pieces of equipment. With a scanner, they produce an exact image of the received frequency that is twice the IF up or down the frequency spectrum. For example, if you are monitoring the National Weather Service at 162.55 MHz and you also note them at 184.15 MHz, your IF image is 21.6 MHz above (or below) the actual frequency because your IF is 10.8 MHz. Image rejection is a statistic that is seldom provided by manufacturers, because most scanners have trouble rejecting this false image. We can use this particular characteristic of

scanners to extend their effective frequency coverage, however, and this will be discussed later in the manual. More advanced scanners use a technique known as *up-conversion*, which puts the IF at around 610 MHz. This is an excellent feature to have when you are conducting high-speed search/scan operations.

4. *Memory capacity.* The monitoring hobbyist may find that a 10-channel scanner is more than adequate to monitor his local police and fire department frequencies. The UW interceptor will need a several hundred frequency capability to monitor all the different agencies and channels that may be in the target region.

5. *Scan speed.* Scan speed is important, and ideally should be selectable from the keyboard. Some scanners will scan at a rate of two to five channels per second, while others will scan anywhere from eight to fifty channels per second.

6. *Search capability.* Most scanners are keyboard programmable, but they will only scan known frequencies. This is inadequate for complete radio-spectrum monitoring. *Search mode* is a program that instructs the scanner to search every frequency within user-selectable limits that may or may not be in use by the target agency. The search feature allows the operator to locate unusual frequencies and then enter them into the scanner's memory for constant monitoring. At least one of your receivers must have a search feature.

7. *Scan delay.* Radio traffic is between two or more persons. If the scanner has a delay feature, it will wait a few seconds before it continues scanning to allow the other party or parties to respond to the original call. Not all scanners have the delay feature. Other scanners have selectable scan delay, which is ideal because it gives you the capability to select the amount of time delay you want

for a specific frequency.

8. *Lock out.* Virtually all scanners have this capability, which allows the user to remove a channel in the memory from being scanned. If a certain frequency keeps locking up your receiver, then lock out can be used until you have time to program in another, more desirable channel.

9. *Monitor or manual.* This feature allows you to stop at a certain frequency and continuously monitor that channel. It is useful for times when your target has secured that channel for a specific operation, or when you find a short but suspicious transmission that you want to investigate. The monitor function can be assigned to a multiband portable analog receiver or to a 10-channel programmable scanner so that your search and high-capacity memory units can be used to continuously locate other channels or monitor other target frequencies.

10. *Priority.* This function assigns a channel priority for a two-second sampling.

All of the scanners we will be discussing are of the programmable frequency, synthesized variety. Crystal scanners are also available, but they serve no useful purpose for the intercept station. Other desirable features for a scanner include compatible power, sensitivity, variable IF, and modulated carrier squelch. These will be discussed as we review specific pieces of equipment.

Record capability is also important. An earphone jack on the unit can serve this function, although some of the better scanners have a *record out* jack for this purpose.

Your station will require more than one scanner for efficient operation. One unit will be assigned continuous search operations. Another will function as a high-capacity memory device for continuous scan of several hundred frequencies that were previously located in the search mode. Another will be assigned a priority mode of ten or so channels that you want to pay particular attention to. A portable, multiband scanner may also be useful in your station for dedicated monitoring and extremely high-speed searches of several bands.

Another type of scanning receiver that is vital for your intercept site is a high-frequency communications receiver. As you will note in Appendix B, most military and government agencies employ HF radio for long-range traffic. Since most scanners deal with FM, VHF, and UHF radio frequencies, a dedicated HF receiver can be quite useful, particularly if you are intercepting traffic from the Central Intelligence Agency (and foreign intelligence agencies), Federal Bureau of Investigation, U.S. Secret Service, U.S. Drug Enforcement Agency, U.S. Customs, INTERPOL, foreign or domestic military services, or certain covert guerrilla operations.

These and other agencies use HF for secure worldwide radio traffic. A dedicated receiver for local scanning of their frequencies doesn't need to be overly sophisticated or sensitive, but it should at least be able to scan a variety of frequencies or bands.

Other considerations regarding receiver equipment include the availability of spare parts, service and technical support, modification capability, and cost. The following table lists available equipment, features, prices, and sources.

High-Capacity Portable UHF/VHF Scanners

Model	Frequency coverage (MHz)	Speed	Memories	Price	Sources
ICOM R-7000	25-1000, 7 CPS 1025-2000	100	$1,020	EEB, HS, EC,	GE, GRV
YAESU FRG-9600	60-905	16 CPS	99	$530	EEB, HS, EC, GRV
PRO-2004	25-512, 760-1300*	16 CPS	400*	$420	RS, GRV
PRO-2021	30-54, 108-136, 138-174, 380-512	8 CPS	200	$330	RS
PRO-34	30-54, 108-136, 138-174, 380-512, 809-906*	8 CPS	200	$330	RS

Model	Frequency coverage (MHz)	Speed	Memories	Price	Sources
BC-205XLT	29-54, 118-175, 406-512, 806-956*	16 CPS	100	$290	SW, GE, GRV
BC-950XLT	29-54, 118-175, 406-512, 806-956*	15 CPS	100	$290	SW, GE, GRV
BC-760XLT	29-54, 118-175, 406-512, 806-952*	15 CPS	100	$280	GRV, SW, GE
BC-800XLT	29-54, 118-174, 406-512, 806-912*	15 CPS	40	$260	GRV, GE, SW
BC-200XLT	29-54, 118-174, 406-512, 806-960	16 CPS	200	$280	GRV, SW, GE
Regency TS2	29-54, 118-174, 406-512, 806-950	40 CPS	75	$290	SW, GE
Cobra SR-15	29-54, 108-174,	8 CPS	100	$230	SW, GE

Model	Frequency coverage (MHz)	Speed	Memories	Price	Sources
AR-900	27-54, 108-174, 406-512, 800-950	13 CPS	100	$260	SW, GE, GRV, ACE
AR-800	30-50, 118-175, 436-526, 800-1000	13 CPS	20	$250	SW, GE GRV, ACE
SR-1000	100 kHz-1000 MHz	40 CPS	1,024	$3,000	GRV

* These frequencies come from the factory with blocking circuitry to stop 800 MHz cellular phone monitoring. This capability is easily restored by simply clipping or replacing a diode.

Sources: ACE=Ace Communications; EC=Electronic Center; EEB=Electronic Equipment Bank; GE=Galaxy Electronics; GRV=Grove; HS=Ham Shack; RS=Radio Shack; SW=Scanner World USA

Ratings (in descending order based on manufacturer's specifications and lab tests):

Computer controlled—ICOM R-7000, YAESU FRG-9600, SR-1000
Sensitivity—SR-1000, ICOM R-7000, YAESU FRG-9600, Regency TS2, BC-950XLT
Scan speed—SR-1000, Regency TS2, PRO-2004, BC series
Image rejection—SR-1000, ICOM R-7000, PRO-2004 (all have *up conversion*)
Frequency coverage—SR-1000, ICOM R-7000, PRO-2004, YAESU FRG-9600
Memory Capacity—SR-1000, PRO-2004, BC-205XLT, PRO-34

SCANNER NOTES AND MODIFICATIONS

The equipment listed in the scanner table has specific applications that are useful for intercept operations. Other devices are commercially available, but these are considered to be the top fifteen units on the market. Other features that might be considered are CTCSS (Continuous Tone Coded Squelch System) tone-decoding options (available on the BC-760XLT), which provides the interesting capability of determining the target's subaudible-tone transmission. This is useful in repeater jamming. If you need to intercept government and satellite transmissions, special antennas and 180-420 MHz coverage are available for the ICOM R-7000, YAESU FRG-9600, PR0-2004, and SR-1000 units.

Basically, you will need to have the broadest frequency coverage and the highest memory capacity available. As discussed earlier, the most efficient intercept listening post should have several scanning receivers on line, since the following operations should be continuously performed by your monitor team:

1. Search/scan of specific bands for covert frequencies.
2. Priority monitoring of five to ten frequencies.
3. Tape recording of one or more frequencies.
4. Continuous scan of 100 to 400 frequencies.

These technical needs can be met with four or more scanning receivers. Priority monitoring can be accomplished with an inexpensive 10-channel hand-held unit, and recording can often be accomplished using a multiband portable with a voice-activated recorder.

The covert listening post in Figure 6 contains the following equipment for VHF and UHF scanning:

1. PR0-2004 400-channel scanner (modified)
2. PR0-32 200-channel scanner (modified) *

3. PR0-38 10-channel scanner (modified) *
4. SW-60 multiband portable receiver

(* Both of these units have 800 MHz converter modules in line. These small boxes provide expanded coverage to many scanners that don't have the capability to intercept cellular traffic. They connect between the antenna and the input and require no internal circuit modification. Sources for this equipment are listed in Appendix C.)

The above equipment provides continuous search/scan of targeted bands (two can be searched simultaneously), as well as extremely high-speed search using the tuning dial on the multiband portable. Over 600 frequencies can be stored and scanned at one time at a combined rate of 34 channels per second. Using this equipment, the entire usable voice-radio spectrum can be covered in the VHF/UHF radio ranges.

This is a low-cost, high-capacity, sophisticated intercept system that is available from any Radio Shack store for less than $1,000. It would be more than adequate for most of the covert intercept applications discussed in this manual.

The PR0-2004 modifications are quite simple. Remove the unit from its case and locate diode D-513, which provides blocking of the 800 MHz cellular traffic. (It is clearly marked on the circuit board. See the service manual if you have any question.) Remove this diode and place it in socket D-510. The PR0-2004 will now scan 400 channels instead of its standard 300 channels. Be aware, however, that this modification voids your service warranty.

Refer to *Popular Communications*, August 1987 and November 1988, for detailed instructions for this modification. A 400-channel template, extra diode, and complete instructions for this modification are also available from Spark Publications (P.O. Box 851, Port Townsend, WA, 98368).

Warning: This modification is a criminal act in violation of the Electronics Communications Privacy Act (ECPA).

Preamplifier

There are several options available to increase the sensitivity and range of a digital scanner. Most of these devices are simple preamplifiers that connect to the receiver's input before the antenna to increase the effective range of the scanner. You can use a basic TV UHF/VHF in-line preamplifier for this purpose, or a commercially available preamp designed for a scanner. Appendix C lists several sources for these devices.

Spectral Display Unit (SDU)

Most of the better scanning receivers have this option available. A spectral display unit allows you to actually see the signals in a specific bandwidth on a CRT (cathode ray tube) screen or a computer monitor. This feature is expensive, but it can be useful for surveillance.

HF RADIO EQUIPMENT

The frequency range between 2 and 30 MHz is considered to be in the high frequency (HF) radio spectrum. Government and military use of this range is constant (certain government frequencies are noted with the mode in Appendix B). HF radio communications can be in several different modes. Some of the more typical modes are continuous wave (CW), upper sideband (USB), lower sideband (LSB), amplitude modulation (AM), radio teletype (RTTY), and facsimile (FAX).

HF radio has many advantages over the VHF/FM/UHF spectrum. HF can make use of sky-wave propagation and provide the user with long-range worldwide communications. Intelligence agencies and the military send coded radio traffic over the HF radio spectrum because of its reliability and ease of use over long distances. U.S. Army Special Forces operations make extensive use of HF radio for this reason.

Depending on your targeted agencies, you may or may not consider a sophisticated HF receiver. Nonetheless, it is recommended that you have the capability of monitoring HF traffic, so you should consider a less-expensive unit as part of your basic intercept equipment. There are several features that are available on modern HF radio equipment that should be considered:

1. *Frequency range.* Select a receiver that can cover the entire HF radio spectrum from 150 kHz to 30 MHz.

2. *Receiving mode.* Some receivers are simply shortwave units that only pick up AM mode, making them worthless for serious intercept ops. Ideally, you want mode selectability for CW, USB, LSB, RTTY, AM, FM, and FAX on the receiver.

3. *Channel memories.* Like scanners, newer receivers have from 5 to 200 memories that can store the frequencies you wish to monitor on a regular basis.

4. *Frequency selection.* Most receivers have a tuning dial and possibly a keyboard for direct entry of frequencies, as well as a memory-recall mode that can select frequencies already entered into memory.

5. *Scanning.* Better receivers have several scan modes which, at the touch of a button, enable the operator to scan an entire band, scan all the channels in memory, or search through specific frequency limitations. Scanning is vital to rapid search/scan operations.

6. *Sensitivity.* The better the receiver, the more sensitive it will be rated. This rating is in microvolts, and the lower the rating the better.

7. *Image frequency rejection.* This feature stops the IF image and the harmonic of the target frequency from interfering with reception. It is measured in decibels.

8. *Bandwidth switching.* This highly useful feature is important for SSB and CW monitoring. It allows the user to select various bandwidths, such as wide, intermediate, and narrow.

9. *Filters.* Noise blankers, IF notch filters, and other filters all enhance the signal quality of the receiver's audio output.

10. *BFO.* A beat-frequency oscillator allows you to tune and clarify SSB and CW traffic. It is vital that it be standard on your receiver.

Review several HF receivers before you decide on one specific unit for your listening post. Also, consider power compatibility and antenna requirements. There are other options to consider for your HF equipment, depending on your target traffic. The following devices are available:

1. *Antenna tuner.* This small device allows you to fine-tune your antenna for a specific frequency.

2. *Antenna preamplifier/Active antenna.* These devices provide a substantial increase in the received signal by amplification at the antenna terminal. An active antenna is one designed with an amplifier already installed into the system.

3. *Multimode data reader.* This is a dedicated computer device that decodes an HF receiver's signal. For instance, it can read Morse code and print it out on a screen or computer printer. It can also copy RTTY information as it is sent from INTERPOL, or it can copy FAX signals from military and commercial stations. These devices generally require a sensitive receiver and a good antenna system to be completely effective.

Model	Frequency coverage (MHz)	Memories	CRIS	Price	Sources
JRC NRD525	.09-34	200	yes	$1,200	EEB, GRV, UN
ICOM R 71A	.10-30	32	yes	$900	EEB, GRV, UN
Kenwood R5000	.10-30	100	yes	$850	EEB, GRV, UN
Sony ICF-2010	.15-30	32	no	$350	EEB, GRV, UN
Sangean ATS-803A	.15-30	9	no	$200	EEB, GRV, UN
Realistic DX-440	.15-30	9	no	$200	RS

CRIS = Computer Radio Interface System capable

Sources: EEB=Electronic Equipment Bank; GRV=Grove Enterprises; RS=Radio Shack; UN=Universal Shortwave

Sources for these options are listed in Appendix C. The covert monitor post in Figure 6 is specifically designed for local line-of-sight intercepts, but it also contains a DX-440 digital communications receiver for search/scan operations in the local HF range.

INTERSTATION COMMUNICATION EQUIPMENT

A listening post should have a means of instant communications with team headquarters, the jamming unit, security elements, and the action group. This interstation COMMLINK is vital for the maximum effectiveness of your system.

Monitor personnel should be able to advise other operational elements of target activity, new or alternate jamming frequencies, raid and search warnings, and the like. This is ideally accomplished by telephone wire. Unfortunately, you may not have the tactical ability to maintain telephone access at all your locations, so radios are recommended. Crystal-controlled, hand-held, or programmable walkie-talkies are useful for this application. There are many such devices available for consumer use at low cost, including citizens band radios, 49 MHz walkie-talkies, and marine-band VHF transceivers.

Interstation covert communications require a variety of frequencies and modes. Special Forces communications personnel specialize in this type of radio and can be consulted for setup and deployment of indigenous equipment for a dedicated communications network.

Signal security (SIGSEC) or communications security (COMMSEC) is not based on the frequency used as much as it is on the encryption of the radio traffic. Using normal transceivers and a variety of different codes, you can send secure messages on a regular, reliable basis.

Citizens Band Radio

CB radios are used extensively by business and private parties in the United States and around the world. One consideration in interstation covert communications is the need to have a cover for the easily detectable and receivable radio traffic you are sending. CB allows this, since you can hear just about anything on a typical citizens-band radio. This makes them ideal for interstation communications.

Most modern CB radios are synthesized 40-channel units that sell for less than $100 in department stores. They have a range of about five miles with an inexpensive quarter-wave whip antenna. AM/SSB CB radios are three times as powerful as regular AM CB radios, so they should be considered as well.

An interesting alternative using CB equipment is the older but still available 5-watt crystal-controlled walkie-talkie. There are two methods of providing a degree of privacy with these units.

First, you can purchase crystals for the desired frequency from Radio Shack, but instead of getting only one frequency you can get two frequencies. Put one crystal in the transmit slot on one frequency and another crystal in the receive slot on another frequency. This allows you and one other station to conduct a conversation on two different frequencies. Of course you may only use two walkie-talkies for this application.

The second alternative is to place one of the remote-control radio frequencies in the desired channel crystal slots. These frequencies (27.195 MHz, 27.145 MHz, 27.095 MHz, 27.045 MHZ, and 26.995 MHz) are used for door openers and car-alarm pagers and cannot be picked up on standard CB radios.

* * * * * * *

As we have learned, there is no such thing as a secure radio frequency. Although you may have bootleg radio equipment, your transmissions *will* be monitored. You may have privacy if you are on an illegal frequency, but your entire operation could be compromised if you are detected where you don't belong in the radio spectrum. If this occurs, all the OPSEC codes in the world won't help you.

Marine- and business-band radio equipment can be used if you have a way to legitimize your traffic. Ham radio equipment is inexpensive and can be considered as well, but again, if you have no license, your operation may be compromised.

The monitor station in Figure 6 contains an AM/SSB transceiver, a 40-channel AM base, and two hand-held portable units for covert crystal use. None of the equipment is illegal to own and operate. Nonetheless, you should keep all unauthorized crystals out of the radios while they are not in use. Your entire station will be confiscated, and you may be fined or arrested if you are caught alternating channels or using unauthorized, remote frequencies.

Chapter Five

Covert Antenna Systems

The antenna is the most crucial component of a radio system. The better tuned and higher up the antenna is, the better its signal reception. Covert placement of improvised antennas is vital to your intercept station's OPSEC.

Most of the signals that we will focus on—FM, VHF, and UHF traffic—have unusual characteristics that require careful consideration. First of all, this type of radio traffic is almost always line-of-sight, meaning the transmitter or repeater you will be monitoring is usually within ten to thirty miles from your intercept station. The intercept site should be as close as possible to your target or targets. (Specialized equipment for longer-range radio intercept on these bands will be discussed later in the manual.)

There is a variety of scanner and receiver antennas available for consumer use. The amateur hobbyist has the luxury of placing his reception equipment on a metal tower in his backyard or on a mast on the roof, but the system we will develop must be invisible to the casual or scrutinizing observer. Fortunately, signal strength is not significantly hampered when antennas are placed in an attic or in the highest point in the radio room. There are also several antenna choices that resemble TV antennas or car radio aerials that are reliable options for covert intercept operations.

DESIGN CRITERIA AND PLACEMENT OF INTERCEPT ANTENNAS

There is no such thing as the perfect antenna, as each design has limitations. Yet during World War II, spies used crude but effective long-wire antenna systems strung along attic rafters or looped around the ceiling in their covert radio rooms to deter observation. We can do the same. The intercept site will have several antennas at the location, and careful placement and a degree of experimentation is required in order to maximize the entire system's efficiency.

There are five basic antenna designs that lend themselves to a high degree of concealability and sensitivity. These are:

1. *Wire antennas.* These are vital for HF intercepts. They can be installed indoors to limit outside view, or outdoors using extremely thin wire or cable that resembles telephone, cable TV, or electric utility wires.

2. *Ground-plane antennas.* These omnidirectional, highly efficient designs can be installed in an attic or directly in the radio room for excellent broad-frequency coverage.

3. *Single-element omni antennas.* These devices are easy to fabricate. They can be inexpensive telescopic whip antennas from a consumer product, a homemade dipole, or a custom-made unit designed specifically for the targeted frequency ranges.

4. *Multi-element discone antennas.* These are professional-grade intercept antennas that also can be used

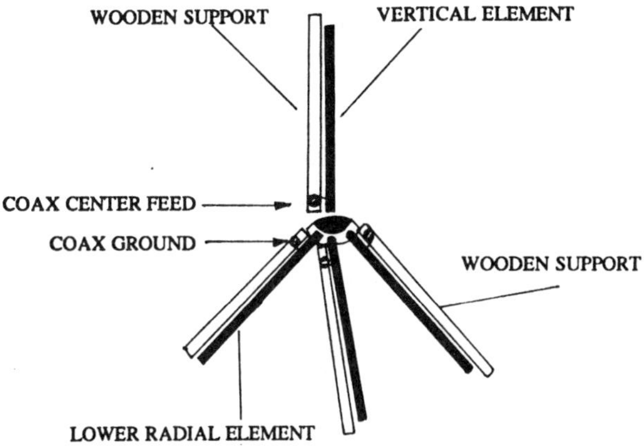

QUARTER-WAVE ELEMENT LENGTH CHART
(234/frequency x 12 = length in inches)

Frequency (MHz)	Length
108	26 inches
110	25 inches
112	25 inches
114	24 inches
116	24 inches
118	24 inches
120	23 inches
122	23 inches
124	23 inches
126	22 inches
128	22 inches
130	22 inches
140	20 inches
------nominal length------	
142	20 inches
145	19 inches
148	19 inches
150	19 inches
155	18 inches
158	18 inches
160	17 inches
162	17 inches
165	17 inches
168	17 inches
170	16 inches
172	16 inches
174	16 inches

Figure 7. An improvised VHF-high ground-plane antenna (108-174 MHz) made of wire and wooden supports. It could also be made with aluminum rods, coat-hanger wire, etc.

for transmitting on typical jamming frequencies. These devices are highly efficient and do not require a large amount of space.

5. *Multi-element beam antennas*. These are perhaps the ideal directional multifrequency intercept antennas. They can be easily built at home or purchased specifically for the target frequency range. The standard antenna seen on most roofs is a highly efficient beam antenna tuned for

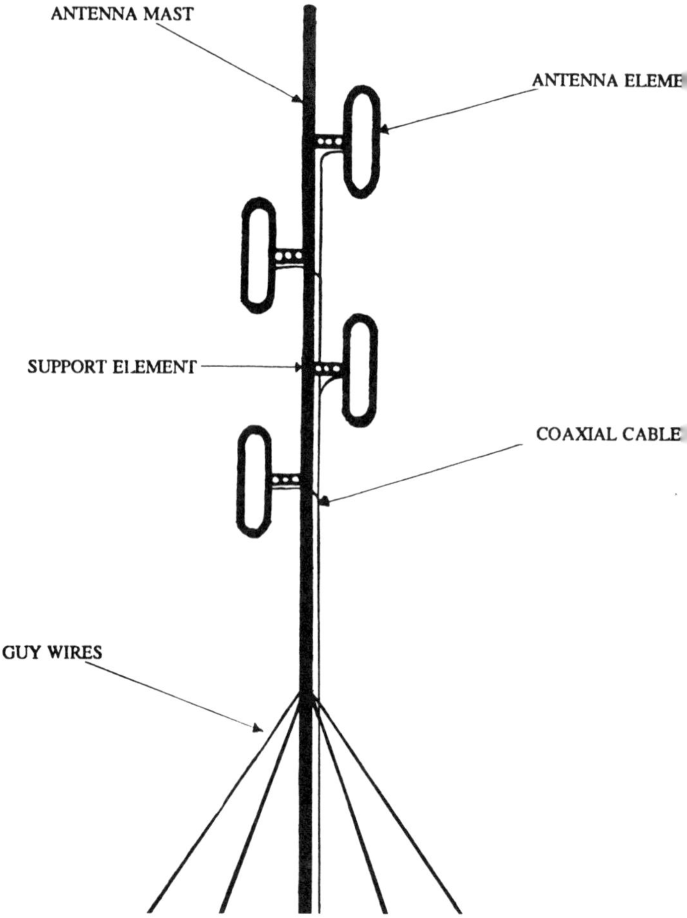

Figure 8. A four-element UHF base station antenna (450-512 MHz).

the TV-signal spectrum. Most beam antennas can also be used to transmit on target frequencies.

Improvised antennas in the above configurations are inexpensive and highly effective for our desired applications. Quad Yagi, loop, and remote active antennas have some use as well, but we will restrict our covert antennas to the above types.

Antennas can be placed in several different areas of your listening post. A small mast can be installed in the attic, perhaps with a rotor for radio direction finding (RDF). Attic installations, when possible, are ideal for many reasons. Installation, adjustments, and modifications can be done unseen and during bad weather or darkness. The proximity to the receivers lessens the need for signal-hungry coaxial cable, which can cause substantial signal losses at longer lengths. Finally, the covert attic site allows for rapid takedown in case of a security breach.

An antenna should be polarized to match the signal pattern of the radio traffic that you want to intercept. Radio signals are polarized either vertically or horizontally. A police car's mobile radio, for instance, is vertically polarized, meaning the antenna goes straight up in the air.

To intercept or jam a vertically polarized radio system, your antenna should also be vertically polarized. Television, FM, and HF radio traffic is usually horizontally polarized—this is why a TV antenna has several elements horizontal to the roof. Most military and government FM radio traffic is vertically polarized, so your intercept antenna should be vertically polarized if you will be intercepting such traffic.

IMPROVISED WIRE ANTENNAS

The most versatile antennas are made from wire. Wire antennas are the best field-expedient devices because of their design and speed of deployment. A U.S. Army Special Forces Communications NCO (MOS 18E) generally is an expert in the use and deployment of improvised wire antennas.

HF, FM, VHF, and UHF antennas can be configured using ordinary wire, including insulated hookup wire, military-grade telephone cable, coaxial cable, metal clothesline, and even picture-hanging wire. The gauge of the wire is relatively unimportant. Insulated wire is ideal, since it can be safely strung through trees. Thick wire (such as coat-hanger wire) is useful for UHF work because it is stiff yet pliable enough to be shaped into specific designs.

Long-Wire Antenna

This antenna offers a degree of directional capability and is the simplest to make. The long wire can be any length for general purpose HF work; however, the ideal length would be either one-quarter wave, one-half wave, three-quarter wave, full wave, or several wavelengths for optimum results. The length of a long wire can be determined by dividing the desired frequency into 468 for length in feet (or dividing the desired frequency into 142 for the length in meters). The best long-wire antenna for HF and FM work is at least three wavelengths. The wire itself should be strung as high as possible, and it should be strung in the direction of the targeted communications.

Dipole Antenna

This type of antenna is the most common for HF use. Dipoles can be used for virtually any frequency. They are known in military communications as center-fed hertz (doublet) antennas. A dipole is made from two quarter-wave sections of wire, being a half-wave in total length. Divide the desired frequency into 234 for the quarter-wave length (multiply this number by 12 to get the desired length in inches for VHF and UHF frequencies). For general purpose HF work, a dipole should be one-quarter wavelength above the ground.

Rhombic and Half-Rhombic Wire Antennas

These antennas have the greatest directional capabilities and are perhaps the most efficient of all wire antennas. The rhombic antenna is diamond-shaped and is directed toward the desired station. A half-rhombic, or rhombic V, is a V-shaped antenna directed toward the desired station. Rhombics are usually too large for covert applications.

Horizontal-Loop Antenna

When there is limited space for a full- or half-wave antenna installation, this variation on the long wire can be used. String your long wire along the tops of the walls in the radio room, wrapping around the room as many times as necessary to provide a full wavelength of coverage.

Vertical-Loop Antenna

This is a directional antenna that has applications in the UHF and long-wave spectrum. The standard UHF wire antenna on the back of a TV is a vertical-loop antenna. The diameter of the loop is generally one-quarter wavelength of the desired signal.

OTHER IMPROVISED ANTENNAS

Radio operatives working on covert assignments have used everything from metal bedsprings, window frames, and curtain rods as field-expedient antennas. Other items that can be used include the metal elements on the outside of a telephone, unused wiring from an abandoned building, or an old TV antenna.

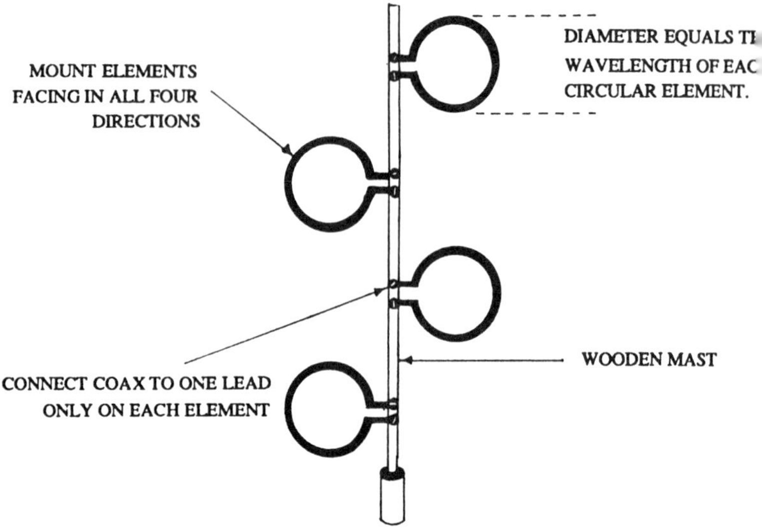

QUARTER-WAVE CIRCULAR ELEMENT LENGTH CHART
(234/frequency x 12 = length in inches)

Frequency (MHz)		Length
390		7.2 inches
400		7.0 inches
	------UHF TV standard element------	
410		6.8 inches
420		6.7 inches
430		6.5 inches
440		6.4 inches
450		6.2 inches
460		6.1 inches
470		5.9 inches
480		5.8 inches
490		5.7 inches
500		5.6 inches
510		5.5 inches
520		5.4 inches
600		4.7 inches
	------nominal length------	
650		4.3 inches
700		4.0 inches
750		3.7 inches
800		3.5 inches
850		3.3 inches
900		3.1 inches
950		2.9 inches
1000		2.8 inches

Figure 9. An improvised omnidirectional UHF four-element circular wire antenna (390-1000 MHz). This can be configured using hook-up wire coat-hanger wire, bailing wire, or standard UHF television antennas.

The following improvised devices can be used in place of wire for FM, VHF, and UHF applications. Practice and experimentation in the construction, deployment, and use of these and all wire antennas are perhaps the operative's greatest skill for a guerrilla operation and the team's most useful radio asset.

Tape-Measure Antenna

A metal tape measure is useful for measuring your improvised wire antennas to a specific length before cutting, but it also has another application. You can use this device as a super-portable rapid-deployment scanner or HF antenna. A 100-foot metal tape measure weighs about two pounds and has proven to be an excellent HF long-wire antenna for tactical field use. It is small and can be deployed and taken down very quickly.

To construct a tape-measure antenna, simply pull out the desired length based on the target frequency, sand off a portion of the paint on the underside of the tape, and connect an alligator clip to the cleaned section. Check the connection with an ohmmeter to verify a closed circuit, then connect the device to your receiver input. Use two tape measures to construct a dipole antenna.

Using nylon paracord and a weight, you can make a vertical-beam antenna for citizens band or 10-meter ham transmissions. Connect your antenna halfway between the ideal full wavelength of your transmit frequency (the ideal full wavelength for a CB is 36 feet, so the connection is 18 feet). Lock the tape measure at the desired length and suspend the device from a tree using nylon cord (*do not suspend near power lines*). A vertical-beam antenna is exactly one full wavelength of the desired frequency and should be center fed for best results.

A Stanley tape measure, model #34-500, is inexpensive

and extends to over 100 feet. It is extremely durable, featuring a good locking mechanism and a take-up handle that allows you to rewind the entire tape in about fifteen seconds.

Foil-Tape Antenna

Inexpensive burglar-alarm foil tape has many applications in improvised antenna design, including quick rigs, custom-tuned internal dipoles, and interior window-mount antennas. This tape was often used as part of the alarm system in store windows, and is still available from many electronics stores (Radio Shack model #49-502 is a 120-foot roll). The tape's conducting foil is placed around the edges of each window. If the window is broken, it opens the alarm circuit and sets off the alarm. (This type of window alarm is slowly being replaced by vibration detectors, which are harder for burglars to defeat.)

You can use foil tape to make long-wire HF antennas, HF and VHF dipole antennas, and UHF single-element antennas. A long length of tape wrapped several times around the top of each wall in the radio room can provide excellent HF reception while maintaining an extremely covert antenna profile.

Use the antenna design formulas in the HF section to make a specific frequency-tuned dipole antenna for your scanners and other gear. For the best effect, attach your foil-tape dipole to a sliding glass door or a window. If you mount it on a wall, use a stud finder to avoid putting the system near electrical wiring or metal supports, which will affect your reception.

Another interesting application of burglar-alarm foil tape is to make a capacitive antenna using the existing house wiring as your antenna system. You may remember seeing a plug-in television antenna sold through mail order some years ago. This device plugged into the wall and had a connection for the antenna terminals on the back of the set. These somewhat dangerous antennas were used by apartment dwellers in the late 1960s and used a capacitive effect to make the house wiring an antenna.

By wrapping a length of foil tape around the electrical wire of your receiver close to the wall plug, you can have the same effect. Wrap six inches of tape around the wiring for VHF scanning, two inches for UHF scanning, and twenty-four inches for HF monitoring. It is an inefficient antenna, but it could be of use for mobile operations or as a temporary application while you install the main antenna system. It does work better than the back-of-set or internal whip antennas that generally come with receivers, however.

Single-Element Omni Antenna

Any conductor that is vertically polarized can be used as an antenna. FM radios, televisions, cordless phones, and walkie-talkies all have single element antennas. This type of antenna is useful for covert work because it is small, concealable, and very inexpensive.

A curtain rod or a stiff length of wire can be used as a vertical single-element antenna. Such an antenna is easy to fabricate and is adequate for omnidirectional intercept work. The typical device is one-quarter wavelength of the desired signal.

A commercially made antenna from Grove Enterprises called the Omni works surprisingly well and is very inexpensive. It comes with all the necessary mounting gear and can sit in a corner or attach to any type of mast (even a broomstick in the attic). The entire assembly is only 66 inches long and can receive the entire spectrum of radio scanner capabilities. It is recommended for your priority 5- to 10-channel monitoring scanner.

Portable RDF Antenna

Portability in a multifrequency radio-direction finding (RDF) antenna is vital for mobile pursuit operations. One useful piece of equipment is a standard set of television rabbit-ear antennas. The two telescopic whip antennas are adjustable in length from 10 to 34 inches. This inexpensive device is available in any electronic store and is easily modified for RDF (see Chapter Eight) and multifrequency mobile scanning.

Before modifying a rabbit-ear antenna, you must understand *impedance*. Impedance is a conductor's reactance to electricity. It is measured in ohms and indicated by the letter Z. Normal television flat-lead cable is 300 ohms, but the input impedance for a scanner is 50 ohms. Therefore, a cable of 50 ohms impedance is required to replace the flat pair of wires running from the rabbit-ears antenna.

Open the small plastic casing at the base of the antenna mount with a screwdriver. Using a hot soldering iron and a vacuum pump, separate the terminals connecting the twin lead and carefully clean them of all wire strands and excess solder. Remove the cable and prep a 50-ohm low-loss video cable for soldering (RG-8 can also be used). Feed the prepped cable through the mount assembly and carefully solder each connector wire to one of the whip-antenna terminals from which you've just removed the original cable. Connect a BNC jack on the other end and plug directly into your scanner.

You now have an adjustable wavelength scanner antenna. This device is tuned by telescoping or retracting the individual antenna elements to specific lengths.

DISCONE ANTENNA SYSTEMS

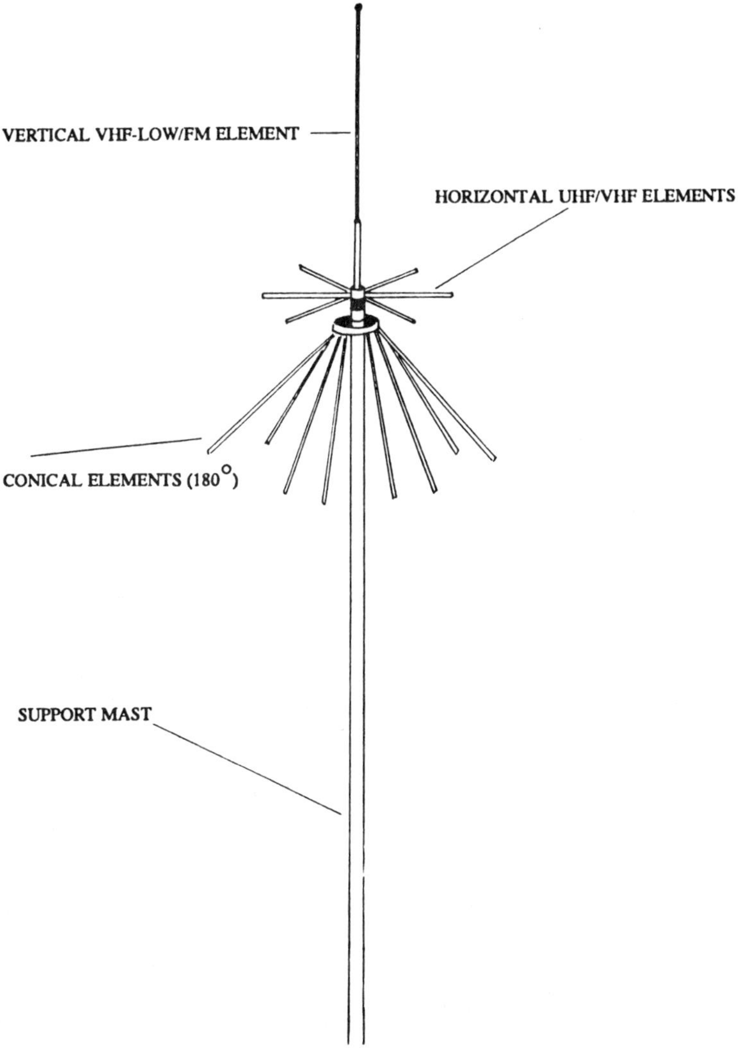

Figure 10. An ICOM model AH-7000 discone antenna with fifteen elements (25-2000 MHz). This omnidirectional, high-gain antenna can transmit and receive on VHF and UHF frequencies.

Discone antennas are sophisticated and omnidirectional, the professional's choice for broad-frequency intercept applications. A typical discone antenna will have twelve to fifteen different elements attached to it. A discone looks somewhat similar to a multi-element ground-plane antenna. Radio Shack model #20-013 and ICOM model #AH7000 are two well-made discone antennas that provide the user with excellent reception on all bands in the VHF/UHF spectrum, as well as transmitting capability on several bands.

The discone is rather difficult to construct as an improvised antenna, but both of the above models can disassemble to a small, space-saving size in a few minutes and are ideal as mainline search/scan antennas.

The discone antenna will out-perform any other omnidirectional antenna and is used by military and government intercept operations for this reason. Intelligence agencies make exclusive use of discone antennas for signal-intelligence applications, and you can spot discones on top of most major embassies throughout the world.

Discone antennas should be concealed in an attic or mounted on a mast in the radio room. It has a distinctive profile, and trained operatives will immediately recognize its purpose. In fact, discone antennas are illegal in some countries. Though their sensitivity and efficiency make them excellent intercept devices, caution is definitely recommended in their use, as they can be very compromising in most areas.

BEAM ANTENNA SYSTEMS

This is the most directional of all antennas. Beam antennas are based on a design configured by Dr. Yagi, who determined that when a dipole was placed near another metal element, a parasitic effect caused a more directional result. These antennas are termed "Yagis" for this reason. A beam antenna can have elements called *reflectors* and *exciters*, which cause a signal to be radiated and received from one specific direction. They can also be connected to a TV rotor for multidirectional scanning and RDF work.

The typical roof-mounted TV antenna is an example of

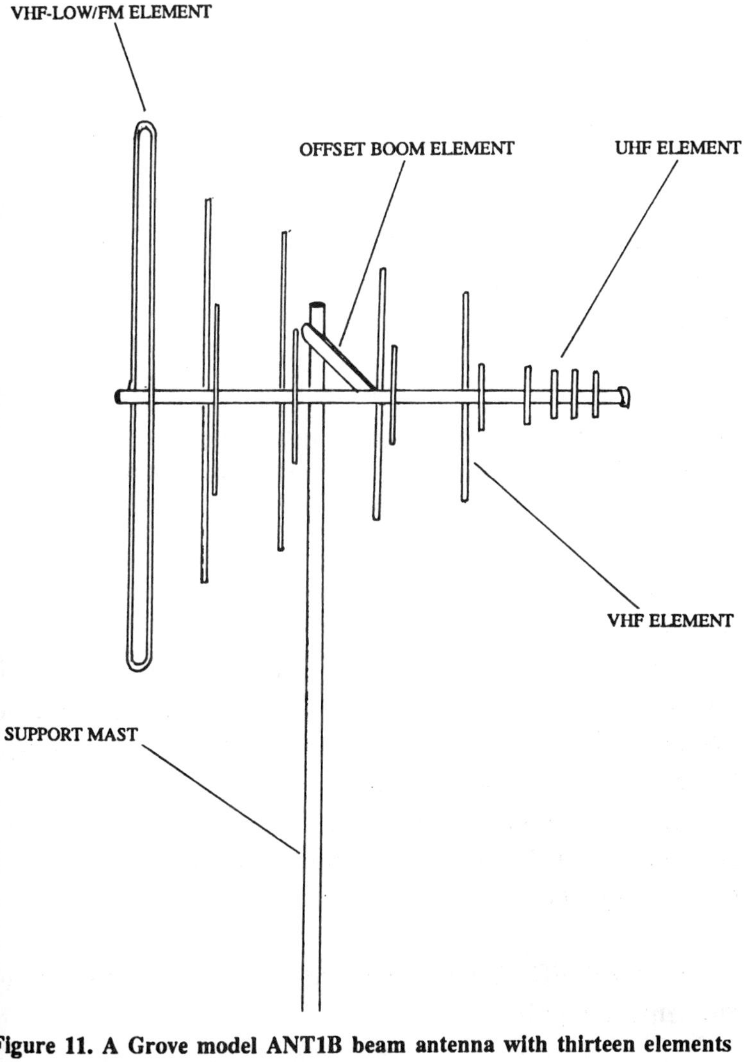

Figure 11. A Grove model ANT1B beam antenna with thirteen elements (30-960 MHz). This directional, high-gain antenna can transmit and receive on VHF-high and UHF frequencies.

a beam antenna. We can modify a TV antenna for covert applications simply by mounting it vertically polarized instead of its normal horizontal polarization. Remove the antenna's mounting brackets and redrill the mounting holes 90 degrees from the current holes. The antenna can then be mounted vertically, making it extremely sensitive for long-range intercepts.

When performing this modification, it is important to match the input impedance of your receiver (usually 50 ohms) to the output impedance of your modified TV antenna (usually 300 ohms) with an impedance-matching transformer (Radio Shack model #15-1253). You may also employ a TV amplifier on the output terminals of the beam for stronger reception from greater distances.

The beam antenna is used by many professional intercept operations. All U.S. Army tactical voice intercept and collection teams use beam antennas so that the intercepted transmission's bearing can be put in the tactical report (TACREP). In an UW environment, two intercept stations operating 90 degrees apart from a target area can use beam antennas for rapid triangulation in locating an enemy transmission or repeater site.

There are several different beam antennas manufactured for scanner work. The Grove Scanner Beam (model #ANT1B) is considered to be the most efficient and sensitive scanner beam antenna currently made.

Although somewhat complex, it is possible to construct an improvised three-element beam antenna using the following formula:

1. DRIVER (center element): 2 one-quarter wave elements in dipole.

2. EXCITER (front element): .97 x one-half wave

spaced .15 x wavelength away.

3. REFLECTOR (rear element): 1.05 x one-half wave spaced .15 x wavelength away.

It is much faster to purchase a beam antenna or modify a TV antenna, but it is good to know the normal dimensions of a typical beam antenna. If you note a beam on the roof or tower of the target agency, you can estimate a target's operating frequency using the above formula.

GROUND-PLANE VHF ANTENNAS

The omnidirectional and very efficient ground-plane antenna is made from several elements. First, a vertical one-quarter or one-half element is mounted on a nonconducting base made of plastic, ceramic, or even wood. Then, three or more radial elements are mounted below the antenna and connected to ground on the receiver. Each radial is one-quarter wavelength long and placed 120 degrees from the other radials. These small, easily concealed antennas can be constructed using curtain rods cut to size or even coat-hanger wire. They are recommended for attic or radio-room installations due to their broad coverage and sensitive omnidirectional signal characteristics.

A ground plane operates in the 108-174 MHz range, and can be used as a VHF or UHF antenna. They are also used extensively in HF and low-frequency FM applications by military and HF ham operators, in the VHF (AM) aircraft frequencies, and by FM police users such as the highway patrol and state troopers. Learn to recognize these antennas and estimate their approximate size to determine their probable operating frequency.

The improvised "jungle antenna" designed and used by Special Forces operators is a variation of a standard ground

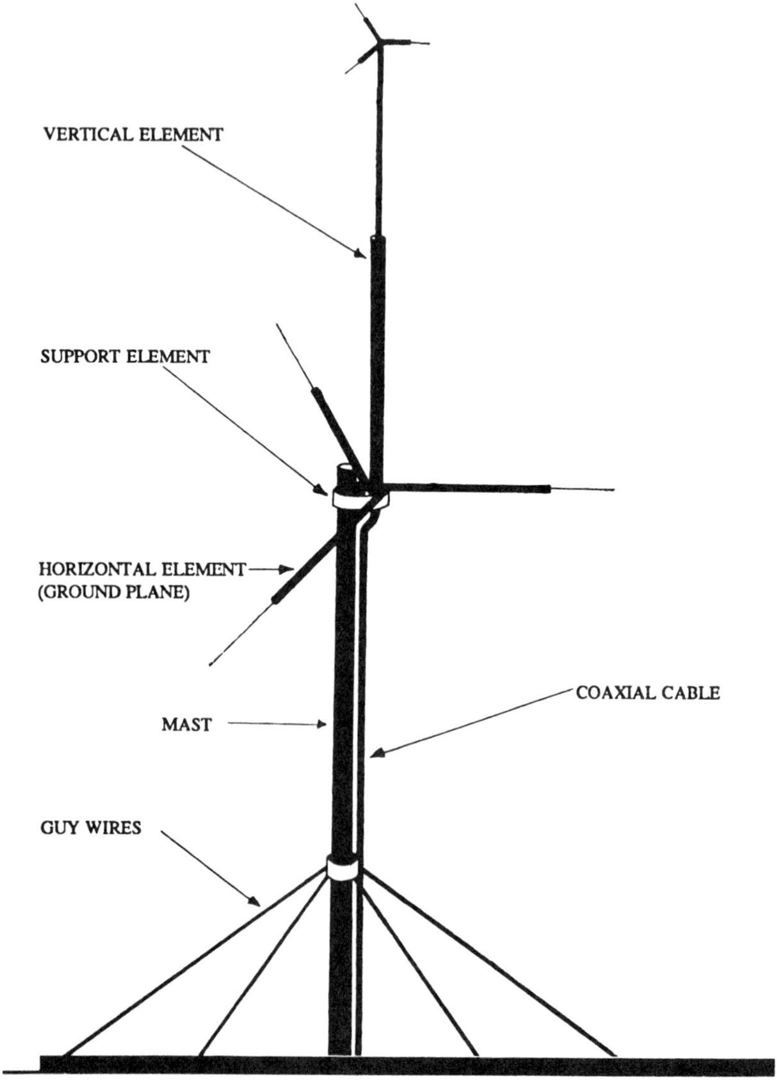

Figure 12. A VHF ground-plane base station antenna (108-174 MHz).

plane. All SF manuals have instructions for designing this efficient antenna. The wire ground-plane antenna described in this manual is based on the SF design. There are also several commercially available ground-plane antennas for your scanner receiver. Radio Shack model #20-176 or #20-014 both work very well.

If your target is at a fixed location, then a ground plane is probably too inefficient. If your target keeps in contact with numerous mobile units operating from random directions, then a ground plane is the ideal choice for an improvised antenna.

VHF-LOW DIPOLE ANTENNAS

These antennas cover the 30-50 MHz frequency range, where you will find most military FM, state highway patrol, short-range SWAT and HRU (Hostage Rescue Unit) communications, surveillance and bugging equipment, and low-power police, fire, and government radio traffic. The signals are either very high-power for wide-area communications or very low-power for close-in communications.

It is relatively easy to intercept these signals with a tuned dipole antenna. Making an FM dipole using coaxial or ordinary wire is not complicated. Cut the wire to the desired length based on the target frequency and place it as high up as possible. Figure 13 illustrates an FM dipole improvised from wire and lists specific lengths for desired frequencies.

An alternative to a VHF-low dipole antenna is a cordless-telephone base antenna. These antennas are tuned to function in the 46-49 MHz band, which is where many of your intercepts will probably occur. This is a second choice, however; the described dipole should be used for maximum signal strength. Note also that the standard "rubber-duck" or back-of-set antennas supplied with most scanners perform poorly in this frequency range.

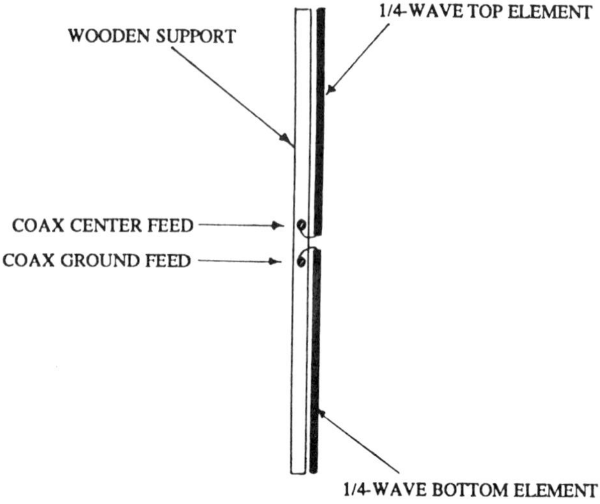

QUARTER-WAVE ELEMENT LENGTH CHART
(234/frequency = length in feet)

Frequency (MHz)	Length
30	7.8 feet
31	7.5 feet
32	7.3 feet
33	7.0 feet
34	6.9 feet
35	6.7 feet
36	6.5 feet
37	6.3 feet
38	6.2 feet
39	6.0 feet
40	5.9 feet
-----nominal length-----	
41	5.7 feet
42	5.6 feet
43	5.4 feet
44	5.3 feet
45	5.2 feet
46	5.1 feet
47	5.0 feet
48	4.9 feet
49	4.8 feet
50	4.7 feet

Figure 13. An improvised VHF-low dipole antenna (30-50 MHz). Elements can be wire, aluminum rods, coax, or even metal curtain rods. Connect elements to feedline (going to the scanner or receiver) at the center point.

HF RADIO ANTENNA SYSTEMS

High-frequency radio operates in a frequency range where horizontally polarized long-wire and dipole antennas work very well. If you have a broad range of frequencies to consider, then a mid-band or commercially manufactured HF antenna should be used. Grove has a 66-foot dipole HF antenna that is highly rated and very easy to install.

The easiest improvised antenna to use on an HF receiver is a simple long-wire antenna consisting of a long length of ordinary wire strung between trees or wrapped around the perimeter of an attic. Using a long wire with an antenna tuner will bring excellent results.

Antenna length is determined by frequency. The following table provides length in feet for specific frequencies that may be targeted for intercept with improvised full-wave long wire, half-wave dipoles, and others:

HF Wire Antenna Formulas
(length in feet)

Targeted Frequency	Full Wave	Half Wave	Quarter Wave
2 MHz	468	234	117
3 MHz	312	156	78
4 MHz	234	117	58.5
5 MHz	187.2	93.6	46.8
6 MHz	156	78	39
7 MHz	133.7	66.8	33.4

Targeted Frequency	Full Wave	Half Wave	Quarter Wave
8 MHz	117	58.5	29.2
9 MHz	104	52	26
10 MHz	93.6	46.8	23.4
11 MHz	85	42.5	21.3
12 MHz	78	39	19.5
13 MHz	72	36	18
14 MHz	66.9	33.4	16.7
15 MHz	62.4	31.2	15.6
16 MHz	58.5	29.3	14.6
17 MHz	55	27.5	13.8
18 MHz	52	26	13
19 MHz	49.3	24.6	12.3
20 MHz	46.8	23.4	11.7
21 MHz	44.6	22.3	11.1
22 MHz	42.5	21.3	10.6
23 MHz	40.6	20.3	10.1

Targeted Frequency	Full Wave	Half Wave	Quarter Wave
24 MHz	39	19.5	9.7
25 MHz	37.4	18.7	9.3
26 MHz	36	18	9
27 MHz	34.6	17.3	8.6
28 MHz	33.4	16.7	8.3
29 MHz	32.2	16.1	8.0
30 MHz	31.2	15.6	7.8

COAXIAL CABLE AND CONNECTORS

All radio equipment requires cables and connectors for power and antenna hookup. A basic understanding of the assorted wires and connector jacks is important for optimum performance of your intercept and jamming gear.

Coaxial cable, or *coax*, is the connection wire between the radio and the antenna. All coax cable has inherent signal-loss characteristics from the transmitter to the antenna and from the antenna to the receiver. It is vital to keep these losses as low as possible, especially in the upper frequencies, where output power is low and receiver sensitivity is affected by significant losses in the cable (also known as *feedline*) running from the antenna.

For HF work (2-30 MHz), the choice of cable is not as critical. RG-8 or RG-58 can be used with good results. Low-loss video cable, such as the type used to connect video cassette recorders and other devices to television and computer monitors, is also adequate. Low-loss video cable has a nominal impedance of 50 ohms, and this is a good match for most input connections to HF receivers. Standard cable TV coaxial cable is generally 75 ohms in the United States and Western Europe—adequate for general coverage HF work but not as good as 50-ohm coax.

The phenomenon by which coaxial cable (or any conductor) causes signal loss is known as *attenuation*. The attenuation of coax is determined by the material it is made of and the outside dimensions of the cable.

For example, RG-8 cable has an outside diameter of .405 inches and is fairly efficient for general purpose radio work. RG-58 has an outside diameter of .195 inches, less than half the size of RG-8. RG-58 attenuates the incoming or outgoing signal considerably more than RG-8 cable. This is not as significant in the HF range, but in the upper frequencies—where we will conduct most of our intercept and jamming ops—signal losses to cable can be considerable.

The best cable for interconnection is Belden 9913 cable. It is efficient but, unfortunately, expensive. RG-8 is the second choice for cable. RG-58 should not be used for most scanner and jamming equipment applications.

There is a bit of a trade-off to consider here. The thicker, heavier, and more expensive cable is the best. Also, the higher up your antenna is, the better the signal. Unfortunately, signal losses in the feedline coaxial cable increases when it is strung high in the air. In the UHF ranges, the signal loss may be so great that a simple whip antenna appears to be more sensitive than a beam or discone in the attic connected to 50 or 100 feet of coax!

Basically, without overemphasizing the technical aspects of feedline attenuation vs. antenna gain, stick with these two rules:

1. Use the best cable you can find and afford.
2. Keep all feedlines as short as possible.

Consumer-grade radio equipment offers a variety of connector jacks for hooking an antenna to a radio. All of these connectors work well to some degree; however, many should be replaced with better adapters when possible.

Like the coaxial cable in the antenna feedline, the connector is also a point where there is substantial signal attenuation. Some connectors are more efficient than others at various frequencies. The following is a list of the most common antenna connectors for consumer- and commercial-grade radio equipment.

1. *RCA*. This is the standard audio or video jack that connects stereo components, computer video monitors, and the like. You will also find this jack on several portable shortwave receivers and hand-held citizens-band transceivers. This connector is extremely inefficient for RF work, and should either be replaced or connected with an adapter to a PL-259 or BNC plug.

2. *Motorola*. This jack is only slightly better than the RCA jack. It is the connector found on car radio antennas and mobile scanners. It is adequate for mobile use, but if a base system is hooked up, an adapter to a PL-259 or BNC is recommended.

3. *PL-259*. This is the standard radio interconnection jack. It can be found on the back of most CB radios and other transceivers, from land-mobile radios to cellular car phones. This jack does cause substantial signal loss in the UHF ranges, but is a definite improvement over the RCA and Motorola connectors.

4. *BNC*. This connector is an improvement over the PL-259 because it maintains a constant impedance over a broad range of radio frequencies. It is the standard for hand-held walkie-talkies and portable scanners. BNC con-

nectors are highly efficient and inexpensive, which makes them the ideal general-purpose RF jack, particularly for UHF work.

5. *Type N*. This connector is found on high-power VHF and UHF antenna terminals and is probably the best connector available. Type N is expensive, but it is superior to the BNC because it is weatherproof.

6. *Type F*. This is the standard UHF/VHF coax connector used for cable TV installations. Type F generally connects to 75-ohm cable and is useful when using modified TV antennas for vertical-beam applications.

To simplify things, use BNC or PL-259 connectors as your standard. Radio Shack has a variety of adapters that will make any plug compatible with virtually any jack. It is equally important to make sure all connectors are as tight as possible, and that the coax has no kinks or breaks in it. If your installation is outdoors or in a high-humidity environment, then tape all connectors once they are tightened and secured.

Chapter Six

Intercept Operations

The covert listening post described in this manual is relatively low cost and quite sophisticated in its intercept capabilities. It can provide your team with valuable intelligence on your target's activities, response and reaction characteristics, and typical patrol patterns.

TACTICAL ADVANTAGES OF INTERCEPT OPERATIONS

There are numerous advantages in having the capability to monitor radio traffic in your area of operations. By continuously monitoring and logging your target's traffic, you can identify its behavior to an almost predictable level. You will know the areas in your target region where the enemy focuses its units and what time of day they appear to be the most active. Within a few weeks of regular monitoring, you will recognize patrol elements by voice and unit number, and often even by name, since first names are regularly used during routine traffic. You will be able to sector the various patrol elements on a map, including where they are typically located. By monitoring their assignments and arrival times to specific types of calls, you will be able to estimate typical response time. You will note that there are certain terrain conditions or obstructions in the area that prevent the units from contacting the base repeater (these are known as *windows* in radar and radio communications terminology). You can also use monitoring to determine training proficiency, unit discipline, and overall morale.

You might even notice certain unauthorized radio traffic from time to time, such as joking and keying the mike to radio music, or accidental keying of walkie-talkies, all of which ties up the base repeater. This is known as a *mike-keyed* condition. You might also notice strange sounds coming from your receiver when two or more units attempt to call in at the same time. This is known as *units-doubling* condition.

With the objective of disabling the target network at will, the UW team must have a complete understanding of the enemy's use of radio. By gathering all of this information, you will be able to develop an intimate knowledge of your target's operational techniques. There is no operation that cannot benefit by monitoring the opposition. You can reduce the chances of detection and apprehension or the chance of being attacked as well as maximize the success rate of a specific planned action.

The UW commander has specific intercept needs in order to assault the target on selected ground. The monitor personnel should be cognizant of these needs. It is important to understand that an efficient listening post will have many uses other than just monitoring a target's radio system. The monitor site described in this manual also provides for short-range covert communications between the site and the action team for constant updates and rapid warning regarding enemy movements.

Tactical voice intercept (TVI) procedures are commonly referred to as *gisting*. The U.S. military defines this activity as providing command with readily exploitable combat information that is passed from the collector to the user as quickly as possible. U.S. Army TC 30-33 describes gisting as "the most efficient, productive method of utilizing tactical voice intercept resources to successfully accomplish the assigned mission." Gisting is basically a handwritten or verbal summary of enemy activities based on their intercepted radio traffic. The gister, or collector, understands the enemy's language, inflections, and slang expressions, and he understands the standard radio codes used to define situations or conditions. The gister learns this information by carefully monitoring and logging the target radio traffic over a period of time.

THE TACREP

A *tactical report* (TACREP) is a standardized form prepared on critical enemy voice intercepts for immediate action and future analysis. It is prepared and sent as quickly as possible—in fact, the military standard for TACREP send-time from intercept to user is less than ten minutes.

Each operation requires a specific set of details in a TACREP, but the following data is generally found in a typical tactical-intercept report:

1. DATE.
2. O/S (operator code).
3. TIME UP/TIME DOWN (when logging began and ended).
4. INTCP UNIT/TEAM (unit and team identification).
5. TAPE # (cassette tape number).
6. TRACK/SIDE/CUT (side of tape and tape-counter location).

7. UNIT IDENT (target unit's identification number or numbers).
8. FREQUENCY (target's frequency).
9. MODULATION (AM, FM wide, or FM narrow; also note if scrambled).
10. ANTENNA (if using beam, give bearing or location if known).
11. XMTR/EQUIP (transmitter equipment; if known, state if hand-held, mobile, or base).
12. CHANNEL (if multichannel only).
13. PAGE (page number of this day's log).
14. CALL SIGNS/PLACE-NAMES (intercepted call signs, locations, etc.).
15. REMARKS (interceptor's comments, notes, etc.).
16. TIME/TO/FROM (note each transmission's time and length).
17. TEXT (hand log of traffic content; gist of conversation).

A TACREP is assigned priority codes that reveal contents and determine dissemination. The military codes for a TACREP are as follows:

PRIORITY 1: Enemy location and/or direction of movement.
PRIORITY 2: Enemy intentions and capabilities.
PRIORITY 3: Enemy operations (data not immediately exploitable).
PRIORITY 4: Enemy routine communications.

Each operation in a UW environment will require different priorities and standards. What would be routine Priority 4 traffic for one operation may very well be vital to another.

If possible, the TACREP should be encoded and secured. The TACREP is the most dangerous component of intercept operations and is the cause of most of the compromised operations studied for this manual.

SIGNAL INTELLIGENCE-GATHERING OPERATIONS

Signal intelligence (SIGINT) is the intelligence product of tactical voice-intercept operations. SIGINT operations are like any other type of intelligence work. The craft is divided into several areas that make up the cycle of intelligence:

1. *Planning and Direction.* Command must first decide on the desired information and then on the desired intelligence "product" that it hopes will be gleaned from that information. Command must also direct SIGINT assets at a specific target or group of targets.
2. *Collection.* SIGINT assets must focus on collection of the desired target information. The collection is "raw," meaning *all* available traffic is collected and stored. No specific type of data is more important than any other at this point.
3. *Processing.* The collected data is sorted and processed by type and decoded into plain text if codes are used. The raw data is then sorted by subject matter for production and analysis.
4. *Production and Analysis.* The processed raw data is studied and converted to numbers or trends. This is then compiled into different focus areas and situations that can be sent to individuals who specialize in the specifics of the resulting intelligence product. The compiled data is reduced to briefs and reports for command use.
5. *Dissemination.* Command decides where the finished intelligence product will be used. The risks of SIGINT operations is greatest at the dissemination point.

The process of gathering intelligence through radio

intercept is very dangerous. It requires excellent operational security and meticulous planning on the part of all operatives. SIGINT is similar to other types of intelligence operations in that it is based on a slow, methodical, detail-oriented collection of bits and pieces of data, a process that is frequently time-consuming and boring. A high-level CIA employee compared intelligence operations to several thousand individuals all collecting and building with small gray bricks. The end result of the building process may or may not be known, but the mundane task of collecting and building is critical to the process, and each brick is vital.

SIGINT is especially mundane because the collector must sit quietly for hours, scanning, searching, and logging with a pair of headphones in a covert location, waiting for the target to give him some data to enter into the system.

Logging SIGINT products into a TACREP is quite mundane, yet it requires a skilled listener who can glean a maximum amount of information from brief transmissions. Since most of the jamming ops in this manual will focus on disabling military or law-enforcement patrols, there will be a definite method to our intercept activities. After you determine your target's operating frequency, you will attempt to determine his operational capabilities, limitations, personnel, response times, and other vital information by carefully studying the traffic's content.

Let's say, for instance, that you have located the target's main patrol frequency and you are continuously monitoring and taping this traffic. You should make a note of the time as well as the unit involved with the transmission. Make a separate file on each unit heard. Learn to identify and study the unit's voice. Male or female? Old or young? Accent? Knowledgeable or an apparent rookie?

Listen carefully to calls that sound somewhat stressful. How does a unit sound when they are in pursuit? Listen to units that have some sort of authority over other units and make a note of that in the unit's file. Do the units occasionally address each other by name? Make a note of it.

From where do these units report their location? Do they have a standard location where they meet while on patrol? Where do they normally take breaks? When do they change shifts? Do they ever play games or joke on the radio? Do they sound tired or edgy? Do they seem to end up in one section of town more than others? Do they go on a lot of false calls to one area more than others? On what types of calls do they generally send two or more units? In which areas does this happen?

Collect as much information as possible on your target organization regarding personnel and personalities. Within weeks, this data can be processed to determine the number of patrol elements, their interactions, their typical response times, their reactions to specific incidents, and their overall efficiency in patrol operations. Learn the agency's codes and special messages. Learn their manpower restrictions, and what time of day they are most vulnerable to unusual situations.

What gets most police and military patrol personnel killed is the routine nature of their activities. Learn these routines. The mundane nature of a unit's day-to-day operations is its biggest vulnerability. On certain calls at certain times, the patrol unit can get complacent and anticipate what it is going to encounter. Their guard is down, and if something dangerous occurs, they are unlikely to respond quickly enough to survive. Rookies are less experienced, but their behavior is more acute, so they are less likely to be sloppy. The young soldier is usually scared and edgy to the point of hyperalertness. Know the rookies from the veterans—they each have different attitudes and responses to situations.

Within a period of time, you will be able to predict locations, responses, and manpower with a high degree of accuracy. You will be able to pick out the target's locations and chart them on a map. You will be able to determine a patrol's area of operations and estimate its ability to get to a specific area in minutes and seconds.

For each patrol unit, be it a military platoon or a police squad, keep the following basic data in your files:

1. Personnel.
2. Calls (domestic disputes, alarms, etc.).
3. Locations.
4. Shifts.
5. Times (activity or no activity).
6. Multiple-unit operations.
7. Codes, signals, etc.

The covert guerrilla operation is very limited in scope. Because of security considerations and manpower restrictions, SIGINT ops may involve only one to five operatives. Your SIGINT collectors must be both producers and processors, so you must stress accuracy and objectivity. Don't allow speculation—stick only to the facts.

Observation is another aspect of the intelligence process. If you hear a patrol element use an unknown code at a specific location, then perhaps it would be beneficial to go to that location and quietly observe what they are doing. Make notes of special operations or situations. If a serious injury or death is involved, then the situation will probably be reported in the next day's newspaper. All local news-gathering operations use scanners at their city desk to acquire news leads firsthand. Use informal contacts with these elements to learn more about your target.

Photograph the target agency and make note of the antennas and towers on the roof or close by. Make special note of the windows or dead spots in the area where buildings or natural obstructions might render a radio inoperative. Study the motor pool area and make note of the antennas on vehicles. Watch individual patrol elements and make note of hand-held walkie-talkies in use. Determine the brand name if possible, and have your frequency counter handy to learn more about the enemy's frequency allocations. These activities are extremely risky if conducted poorly, but they can contribute immensely to your overall SIGINT operation if done properly.

Chapter Seven

Operational Security

It would be a mistake to believe that the monitor site described in this text could be passed off as a hobbyist's setup. The covert antennas, backup power supplies, modified high-capacity scanner, and illegal transmission equipment would be seriously incriminating, and it would be virtually impossible to explain your intentions if you are caught by the opposition. Consider for a moment what your response would be if you discovered some group intercepting your communications, compiling data on your activities, and gathering the technical capability to defeat your radio communications. It is doubtful that you would stop and consider the civil rights of these individuals.

Tactical voice interception is considered spying and/or terrorist-related by many nations (including many NATO countries), even during times of war. In fact, simple possession of this manual may be considered a capital offense in certain Asian, Middle Eastern, and Soviet block countries. Therefore, meticulous OPSEC is a must to reduce your chances of detection.

OPSEC MODIFICATION FOR SCANNERS

All consumer- and military-grade communications scanners have volatile memory chips that store the desired frequencies in the receiver for sequential scanning. When the power switch is turned off, the receiver still receives a small amount of voltage from the power supply to maintain the memory. A volatile memory of RAM (Random Access Memory) chips will lose all of its stored frequencies if it loses this power. For this reason, most scanners have an internal backup system consisting of an extra battery or an internal capacitor. This system allows the device to maintain memory for a time while you change batteries or disconnect the system for transport or relocation in the listening post.

Although there is no question that it saves time to be able to avoid regularly reprogramming a high-capacity memory scanner, it is also a serious security risk to have these frequencies keyed in if the site should be detected or raided. A simple modification will eliminate this risk.

Open each scanner and install a switch between the battery and the backup circuitry. If it becomes necessary for security reasons to quickly disable the memory, then a flip of the switch will clear out the memories of each unit. (Many new devices have a reset or clear button already inside the battery compartment, on the circuit board, or on the back panel that will perform this function.) Make it directly accessible or run an extension switch from this terminal. In sensitive high-risk ops, keep all memories clear when inactive.

Actual raids on monitor posts that have been used for criminal enterprises illustrate the foolishness of not making this easy modification. These operations were clearly monitoring the raiding agency's radio frequencies and/or federal government traffic, offering incriminating evidence directly to law-enforcement officials by getting caught with the telling frequencies programmed into their scanners. If your opposition is less liberal in handling such matters (and most foreign governments are), this type of security breach could become more than just evidence in court—it could cost lives.

TAPE, DATA, AND PAPER SECURITY

Site OPSEC includes the careful use and storage of data-retrieval media and printout accessories. Cassette tapes and data disks should be backed up, meaning extra copies of each should be made and stored in a secure location away from the site. For rapid destruction of disks, cassettes, and most other magnetic storage media, a tape eraser should be kept on site. Available from audio and computer stores, this simple, inexpensive device is designed to bulk erase audio cassettes and data disks in seconds using electromagnetic currents. Keep your working copies of tapes, software, and data disks in the bulk eraser and turn it on in an emergency or security breach.

Oscillations from computer equipment can be reduced by enclosing your video monitor and CPU case in two or three layers of tinfoil. A better alternative is to purchase data-processing equipment that uses a flat liquid-crystal display (LCD) screen. The emanations are much lower than with a TV video screen.

Typewriter and printing-calculator ribbons should be kept out of the machines and secured when not in use. When the ribbons are used up, they should be completely unraveled and burned. Intelligence operatives consider a typewriter ribbon a useful find when sifting through a target's trash, since everything typed into a computer printer or typewriter can easily be read from the ribbon.

Other operational security precautions at the site location are somewhat obvious. Frequency guides should have no markings in them, and they should be concealed or destroyed if possible. Many alternative and underground publishers offer these guides (see Appendix A), and their very contents are incriminating enough. If you mark your frequency selections in these guides, you are really taking a substantial risk. Maps used for siting and RDF triangulation activities should also be unmarked (they should be destroyed after use). All incriminating magazines and other publications should be placed in a locked box or safe, just as you would store classified documents.

PORTABLE DATA-BANK CALCULATOR

Frequency-logging is critical in order to effectively use intercepted enemy radio traffic, but keeping paper documents showing your logging activity is dangerous. A simple program for search/scan operations that will run on all personal computers is listed in Appendix D. Unfortunately, computers can be overly elaborate, expensive, and unreliable. Instead, a relatively new, inexpensive pocket computer can be utilized for logging purposes if your security is especially at risk.

Most of the traffic you will be intercepting can easily be converted into a single word, a time, and a frequency. This can be further abbreviated down to eight letters and twelve numbers. Using this coding system, a relatively new device called the telephone number Data-Bank calculator can be used to log enemy traffic.

The Data Bank is basically a tiny computer dedicated to organizing phone numbers that is easy and fast to operate. This credit-card-size unit stores about 16K worth of information, or about 150 individual entries. It will automatically alphabetize or sort your entries numerically, and it can search for a specific entry. The Data Bank has a reset button on the back panel that is recessed so it can't accidentally be set off, while remaining very accessible in an emergency. The unit costs between $15 and $20, and it is recommended that you buy several units and extra batteries.

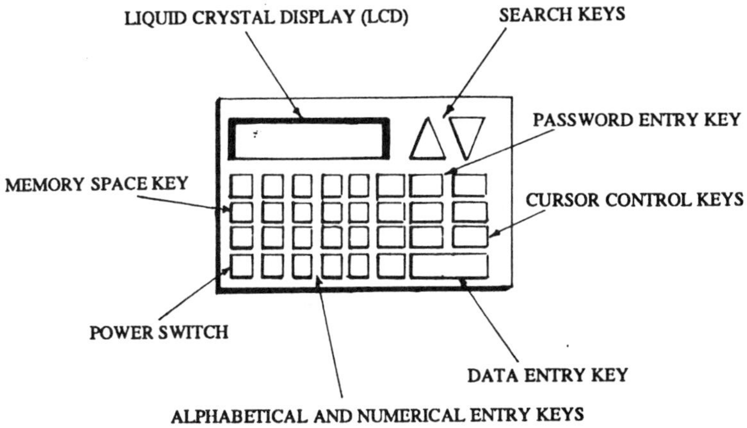

Figure 14. An LCD Data Bank portable calculator. Enter the target's name in the NAME entry (up to eight characters), the intercept time in the AREA CODE entry (up to four numbers), and the frequency in the NUMBER entry (up to eight numbers).

FLASH PAPER AND THE "BURN BAG"

Another security method employed by several intelligence cells is the practice of keeping notes, intercept reports, and frequency data on a flammable paper, commonly referred to as "flash paper." This fiber-bond paper will ignite and disintegrate to ashes in less than a second. Ordinary flash paper available from magic stores and by mail order comes in average-size letter sheets, can be printed or typed on easily, and will "flash" with a touch of a lit cigarette.

An electrically operated "burn bag" can be created by

storing all flash-paper documents in a metal enclosure and hooking up a broken-lens light bulb to a mercury switch and a battery. If the box is tampered with, picked up, or moved, the switch will energize the bulb and the paper will be destroyed in a fraction of a second. This method allows your intercept team to store written documents that can easily and quickly be destroyed in case of a security breach.

PRINTING CALCULATOR AS A SEARCH/SCAN LOG

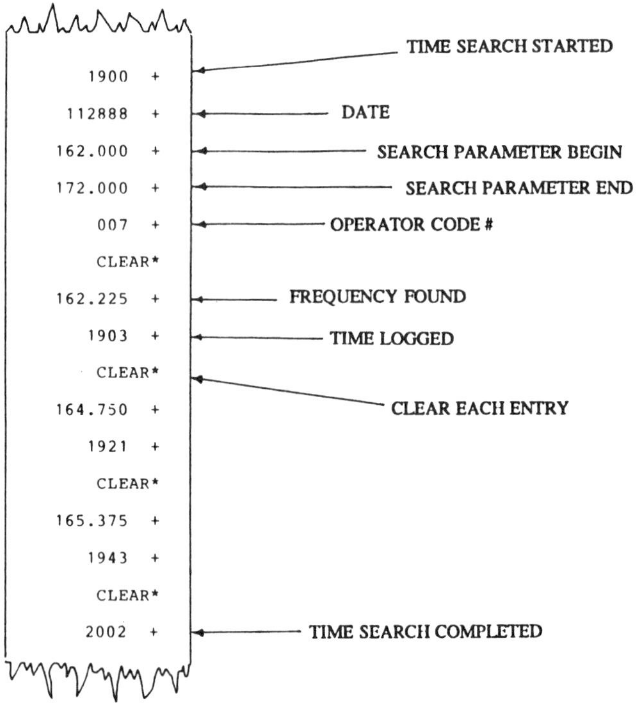

Figure 15. Adding machine tape as a search/scan log. All notes other than those indicated can be assigned numerical codes, such as male or female operator, base or mobile, etc.

When search/scan operations are being conducted (where speed is critical), you will only have a second or two to log a frequency if you are searching as fast as your equipment will technically allow. One useful technique employed by intercept operatives is the use of a portable printing calculator with a paper tape. As soon as the scanner locates a frequency, the operator enters the frequency into the calculator and lets it print. He then types in the hour in 24-hour time on the line below the frequency number. Once activity subsides or the operator is relieved, he can key these frequencies into a small computer for sorting out duplicates, and then into the memory of the high-capacity scanner for future monitoring and ID. Once the data is safely stored, the adding-machine tape can be burned in an ashtray.

BOOK CODE AND BASIC ENCRYPTION TECHNIQUES

A portable printing calculator can also be used as a fast, secure reporting log using a variety of cryptographic codes. This practice appears to be quite common among some intelligence cells. For example, "book code" is a sequence of predetermined numbers in which every six numbers represent a word at a specific location in a book, which is often an obscure dictionary that all parties in the cell have in their possession. The first three numbers of the code represent the page number in the book. The next number represents the appropriate column, and the next two numbers represent the number of words down that column that will bring you to the coded word. By assigning names and code words to your targets, operations, etc., you can have a further degree of crypto-security. This or some other type of code is strongly recommended for intercept TACREPs and other intelligence information that your team must commit to written records.

For additional security using book code for written reports, a sequentially changing alphanumeric code can be used in place of the numbers. The following formula illustrates a sequential assignment for alphanumeric conversion of numbers to letters. There are many advantages to this type of encryption. Primarily, this code allows a degree of controlled randomness that makes it difficult to defeat with computerized code-breaking equipment.

Alphanumeric Encryption

Step One: Convert plain text to number value.

First 3 digits = Dictionary page number.
Next 1 digit = Column number.
Next 2 digits = Number of words down page.

Example:
SUSPEND OPERATIONS AGENT COMPROMISED.
Number value = 299103 198102 008105 056203.

Since all words are exactly the same length, they should be further grouped: 299103198102008105056203, with each line uniform.

Step Two: Convert numbers to letters

1=A, K, or U	5=E, O, or Y
2=B, L, or V	6=F, P, or Z
3=C, M, or W	7=G, Q
4=D, N, or X	8=H, R

9=I, S 0=J, T

Example:
SUSPEND OPERATIONS AGENT COMPROMISED
29910319810208105056203
BSSKJCASHKJBTHAJYTEFVJC

Select conversion letters randomly. Use the second digit of the date to determine where you start the A and count your alphabet down from there. For example, if it is the 23rd of the month, 3 would be A, K, or U; 4 would be B, L, or V; etc.

Encrypting all written communications and using code words for your own radio traffic is vital in this type of operation. Compartmentalize your intercept and jamming units and assign each a method of sending and receiving coded traffic to and from your forward operations base (FOB).

* * * * * * *

Alternate methods of emergency message and data transfer should also be arranged. One improvised method that is extremely fast in emergencies is a method (generally credited to the German Abwehr during World War II) that utilizes an ordinary newspaper and a pin. Punch tiny holes in each letter of your message at a prearranged location in the newspaper text. The message can be quickly read when the newspaper is held up to a light, though it is difficult to otherwise detect.

Another solid security technique uses the Lost and Found section of the newspaper. A prearranged Found advertisement—such as "Abort" or a similar one-word message—can be placed to notify all team elements operating throughout the area of a busted operation.

There are many variations on the above codes and message formats. The bottom line is that OPSEC considerations can not be stressed enough. A commander should implement a personally devised code system for the operation that only he and the operational team will know. Virtually every compromised intercept or jamming operation in history has had operational security as its fatal flaw. It is a major focus area for intelligence by your opposition.

PHASE TWO: TARGET ACQUISITION

Chapter Eight

Antenna Recognition Techniques

Good antenna-systems intelligence can provide your team with sufficient information to neutralize the target's radio system, including his alternate, backup, and tactical radio frequencies. In order to do this, your target must be carefully studied. Use efficient search/scan operations, antenna-length guidelines, frequency counters, and published sources (see Appendix A) to get the whole picture.

Observing the base antenna will help you determine its frequency. Combining this information with observations of marked and unmarked cars and data on walkie-talkies will provide you with a good picture of how your target communicates. Providing this intelligence to your search/scan monitoring team can save a lot of time when it comes to target neutralization. Note that most agencies use a combination of frequencies and bands in their communications, so it is important to make note of *all* antennas.

Failure to learn about even one frequency that your target has access to will cause your entire jamming effort to be in vain. High-risk operations such as "jiggling the wire" (see Chapter Twelve) are unnecessary if you maintain good monitor discipline.

RADIO DIRECTION FINDING

Before you can observe your target's base antenna, you have to locate his transmission site. Radio direction finding (RDF), therefore, is a very sophisticated capability to have. It can be used for locating enemy communications centers and clandestine radio sites, as well as for various search/destroy operations. RDF is also good for zeroing in on specific frequency-compatible radio transceivers to be "requisitioned" for future modification and use as jamming-transmission equipment.

(Terrorist cells seldom build or purchase jamming equipment. They simply steal the necessary radios from a commercial enterprise, such as a taxi company or a utility service, and put the target frequencies into the crystal slots on each unit. It is actually a little more complicated than that, but compatible equipment is almost always available in your target region, and good RDF is the fastest way of finding such gear.)

RDF is not a complex operation. In fact, when you move the rabbit ears on your television for a better signal, adjust your FM radio antenna, or rotate the TV antenna on your roof with your rotor control, you are using the electrical characteristics of the antenna to locate your desired signal—in effect, using radio direction finding.

RDF requires specially tuned antennas and receivers to intercept and determine the direction and location of radio transmission equipment. The FCC has perhaps the most sophisticated RDF capability in the Western world. They have extremely sensitive directional equipment in aircraft and vehicles, and they use "chase" frequencies to conduct searches for jamming or interference transmissions. FCC operations have been quite successful in the past few years. Though your opposition's RDF capabilities may not be quite as sophisticated, you must have excellent timing, high mobility, meticulous OPSEC, and a certain degree of luck to defeat a determined search.

Simple Triangulation Techniques

An important aspect of your intercept station will be the site operator's ability to estimate the direction and/or position of the target's radio traffic. This can be accomplished by several means. The most basic is a technique called *triangulation*, which can be accomplished using an improvised dipole antenna.

Dipole antennas have certain receive and transmit characteristics that can be used for RDF operations. The signal coming toward a dipole antenna is strongest when it is parallel to the antenna elements. By holding your dipole antenna horizontally over your head and rotating it (or by mechanically moving a rotor on a mast) until you get the strongest signal, you can get a partial fix on your target's approximate location. (It is important to note that you use your dipole horizontally polarized for the most effective RDF and vertically polarized for general intercept operations.)

If you have two intercept sites located on the outskirts of your target area, and they are about 90 degrees apart, you can triangulate your target's location by orienting a map and drawing a line from the two (or more) locations toward the direction of the signal. Where the lines intersect is your target's approximate transmission location. A mobile unit, keeping in contact with the base intercept site by radio, can locate enemy stations and their repeater sites. This is a highly useful technique that we will use in our jamming operations. Figure 16 illustrates triangulation on a map grid.

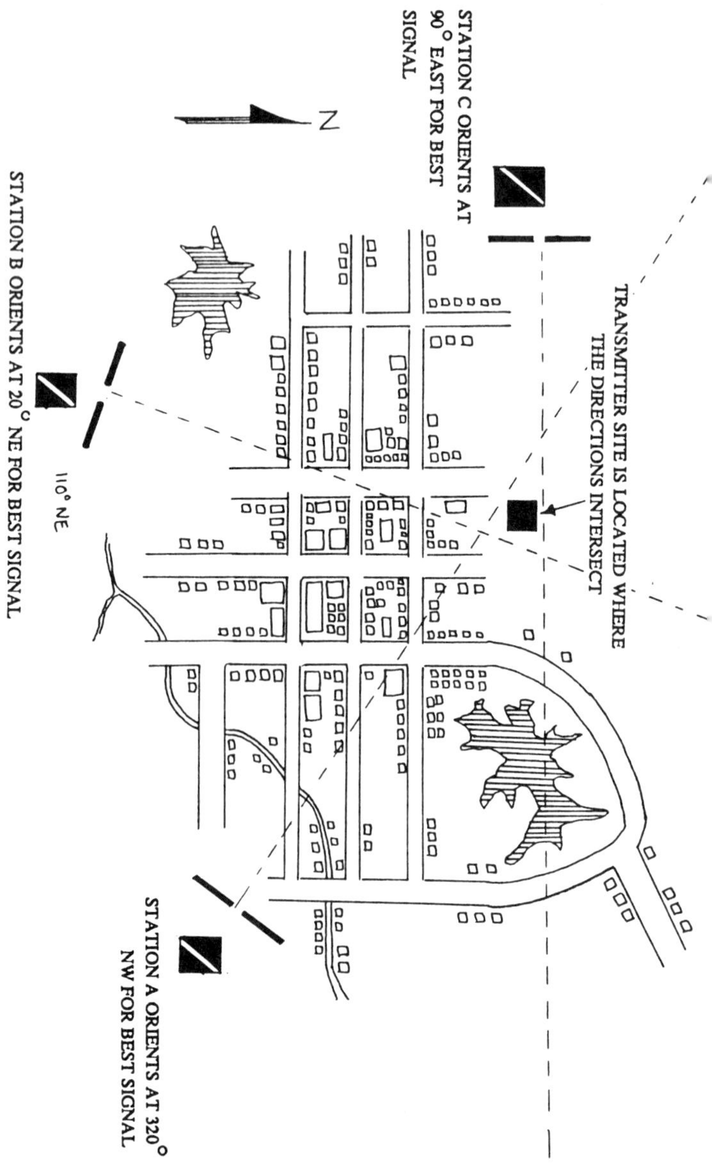

Figure 16. Radio direction finding technique using three monitor stations and direct tuning of dipole antennas. The approximate direction of the best signal from each station is given in compass bearings. The map is then oriented and lines are drawn in the direction of the coordinates.

Most of your RDF work using triangulation will be directed toward locating your target's repeater site. There are some practical limitations in using simple dipole antennas for this work, however. One is that the bearings determined with this method are usually off by 5 to 15 degrees. Thus, the more bearings you have, the better your location estimate will be. This does not mean that you have to employ multiple sites—it is possible to simply go to several locations individually, get a bearing on the target signal, then go 90 degrees or so from that bearing and take another one.

Another limitation of this technique is that FM signals tend to reflect off buildings, hills, and other large obstacles. This can give you false readings. Again, the more bearings from different directions that you get, the closer you will be to locating your target.

Since your target generally will be a large tower or a transmission system on top of a tall building, it should not be that difficult to find. RDF simply accelerates the process and provides you with accurate verification of the repeater's function once it is located. FM/VHF low signals are perhaps the easiest to locate, primarily due to the strength of the repeater signals.

It is interesting to experiment with RDF techniques. The process is simple, and it can be very effective once you have the technique down to a precise operation. Locating intermittent, low-traffic frequencies in other bands is more difficult due to the lower power of the devices and the occasional lack of traffic on the system to assist you in getting bearings. The better you are at RDF, however, the less signal you need to get your bearings quickly.

For truly efficient RDF work, the beam antenna is ideal. The beam has the highest degree of directional sensitivity and is probably the most accurate antenna available to the commercial user. The beam can be mounted on a mast and turned manually, or it can be turned with a motor similar to the kind used for consumer TV antennas. Grove's Scanner Beam antenna has proven to be very accurate in this application.

FIXED-SITE INSTALLATIONS

There are certain normal procedures used by intelligence operatives to study the opposition's radio site. Verification of the purpose and transmission frequencies of the communications installation can be accomplished electronically using search/scan operations and frequency counters. The first step in radio jamming, however, is to identify the target transmitter. Therefore, the size, elevation, orientation, and design characteristics of the target antenna system are vital intelligence information.

The length of an antenna is generally a dead giveaway of its approximate operating frequency. In order to be efficient, different-sized antennas must resonate at specific wavelengths. A bit of practice will train the operative's eye for such details as antenna type. Skill and accuracy are easy to develop in a short period of time.

If the installation is in a denied area, you must rely on either aerial reconnaissance or penetration by a contract agent. Aerial photos of antennas are sometimes helpful, but generally the system will have such a low profile and minimal signature that the most effective way to identify it is by ground-level photography. Since the size of the antenna system is important in determining operating frequency, instruct the operative to note lens-focus settings and an approximate distance when photographing the target. Then you can determine the size of the target by taking a photo of a similar-sized object using the same camera and the same settings and then comparing the print negatives.

This is perhaps an overly sophisticated approach in dealing with a company- or battalion-level military or law-enforcement agency. The typical communication characteristics between patrol elements of such a target pretty much restrict radio traffic to low- and medium-power line-of-sight communications. Chances are you will be able to recognize the target's antenna system profile from one of the systems described in this manual.

Most security-oriented targets will have their radio antennas on top of their headquarters or on a tower within their operations compound. The following observations should be made regarding the site installation:

1. Map location (grid coordinates, etc.).
2. Approximate site elevation.
3. Antenna height and description (sketch or photo).
4. Antenna orientation (if beam type).
5. Number of antenna elements on each mast.
6. Antenna condition (guy wires loose, crooked mast, etc.).
7. Visible site security personnel.
8. Site activity.
9. Frequency-counter loggings while at site.

Obviously, the first consideration in disabling enemy radio capability is the physical destruction of the site. This is generally not a possibility in an urban guerrilla-warfare environment. Access to the site may be possible, but if the team were able to capture the target headquarters, disabling its communications would no longer be necessary.

Well-placed small arms fire can temporarily damage the site installation. This task is perhaps best assigned to a small unit of indigenous personnel. An RPG (rocket-propelled grenade) at the base of the antenna can either damage the coax feedline or drop the antenna completely. Destroying the antenna will disable the radio base; if the target happens to be transmitting at the time, it can seriously damage his transmitter as well.

Gaining access to the site and simply cutting the coaxial feedline would accomplish the same effect, but if accurate small-arms fire can be brought to bear on the antenna, then the coax connection is the desired target, since it will take considerably longer to repair if the damage is up on the tower or mast. (It will also be difficult to convince technicians to climb a tower where small arms fire can be placed so accurately!)

Stringent security around typical jamming targets will most likely preclude the above options. Instead, a penetration team has to provide an accurate description of your target's antenna system. The system is usually visible from a safe distance, and photography is recommended. Note, however, that although site observation seems to be a relatively safe and harmless activity, at least two major terrorist groups have been compromised by alert citizens or patrol officers who observed an operative performing this task. In many countries, photographing any type of government installation is a serious offense.

MOBILE UNITS

Because of the typical target's desire for visibility and its tactical need for constant mobility, observation of vehicle and hand-held antenna systems is generally safe and easy. However, there are still some risks involved with photographing or studying patrol vehicles and personnel, so careful planning is essential to avoid detection. One technique is to set up a "walking surveillance" around tourist attractions in your target area. If properly attired, your operatives will generally go unnoticed if they photograph a popular site that just happens to include a passing patrol vehicle or uniformed officer. You'll also have a certain degree of plausible deniability if you are detained or questioned. Operatives working in foreign countries should take OPSEC into special consideration when performing this task.

Vehicle Antenna Installations

The following observations should be made of patrol vehicles in regard to their radio equipment:

1. Type of vehicle (compact, sedan, four-wheel drive, etc.).
2. Number of antennas and approximate length.
3. Description of each (photo or sketch).
4. Location of each (hood, roof, trunk, bumper).
5. Permanent or temporary installation (clamps or coax cable visible).
6. Frequency-counter readings (if possible—not advised for foreign operations).
7. Walkie-talkies visible (on dash or patrolman's belt).

For typical U.S. law-enforcement targets, there are some rules of thumb that you can use to determine vehicle operating frequencies. These specs applied in about 90 percent of the cases studied:

1. *Highway Patrol/State Police.* Trunk- or bumper-mounted antennas, 4 to 7 feet long, or trunk-mounted with center load (a 3- to 4-inch-long plastic or metal cylinder in the middle of the antenna whip element), approximately 3 feet long. Operating frequency is 30-50 MHz. There probably will be a citizens-band radio antenna and possibly a cellular phone on each unit as well.

2. *County Sheriff/Rural Police Department.* Trunk-mounted antennas, approximately 1 1/2 feet long, or trunk-mounted with bottom load (plastic or metal cylinder located at the base of the antenna), approximately 4 feet long. Operating frequency is 150-174 MHz.

3. *Municipal Police.* Trunk- or roof-mounted antennas, approximately 5 inches long, or center loaded, approximately 3 feet long. Operating frequency is 450-470 MHz. May also have cellular phone system. Cellular or 800 MHz radio antennas may be on any of these vehicles. The antenna is either about 3-4 inches long, or about 1 1/2 feet long with spring center load.

Both undercover operatives and certain federal agencies have an unusual requirement for antenna systems on their mobile units. The vehicle must not look like a law-enforcement vehicle with the usual assortment of protruding antennas. Instead, these agencies use disguised AM/FM radio antennas. There are two ways to accomplish this. One is to use a loading coil under the dashboard that makes the vehicle's standard antenna resonate at the desired frequency. This is a somewhat inefficient method and is not used much anymore.

The other method is to mount a new antenna where a normal AM/FM radio antenna would be on the quarter panel of the vehicle. Outward appearances will not give this antenna's true purpose away, since it looks exactly like a normal radio antenna. Yet there is a telltale sign that you can look for to identify these covert antennas. On late model cars, the antenna often is a hairline copper wire that runs down the middle of the windshield. Municipal police seem to forget this. Since a key element of their undercover work is a constant change of vehicles, they often get their unmarked cars from the open market and simply mount a look-alike radio antenna on the front quarter panel, even though the car normally wouldn't have one. The copper wire on the windshield gives this away.

If your target is a federal agency, avoid attempting to scrutinize suspicious vehicles. These personnel, particularly in the foreign and domestic intelligence profession, are unlikely to provide you with observable technology. Your monitoring is best restricted to search/scan operations from a covert location. Appendix B contains several hundred known frequencies for specific government agencies.

Portable Equipment

All line-of-sight communications for law enforcement and platoon- and squad-level military applications require that each patrol carry portable walkie-talkies when they're on surveillance or away from the vehicle for any reason. These small radios sport antennas that can give away the operating frequency.

Older hand-held radios, particularly VHF low and VHF high radio gear, have a telescopic whip antenna that extends to about 4 feet in length. These more cumbersome antennas are actually much more efficient than the flexible rubber-duck antenna seen on most modern walkie-talkies.

The rubber duck is actually a helical-type antenna. A technical description is not critical here, but it is important to know that the length of these antennas is still a good means of determining their operating frequency. VHF high or VHF low gear generally runs 12 to 18 inches. Anything from 8 to 12 inches can be VHF high or UHF. Rubber duck antennas that are 4 to 6 inches are usually UHF.

FREQUENCY COUNTERS

A frequency counter is a small, digital broadband radio receiver that provides the user with an accurate readout of a received radio frequency. There are several models available on the American and European markets. Prices range from $135 to several thousand dollars, depending on the features, range, and sensitivity of the unit.

A frequency counter can be connected directly to the output of a transmitter, or it can be connected to a small antenna and placed close to the transmitter to provide the transmitter's output frequency. It can also be used to detect and locate hidden transmitter devices such as eavesdropping bugs, telephone taps, and individuals wearing wires. Most importantly, it can determine your target agency's covert transmission frequencies, which makes it vital to have in your radio kit.

The Optoelectronics portable frequency counter (model #1300H) is only 3 1/2 by 4 inches in size, and can easily fit in a jacket pocket. It has a BNC connector jack, which allows a wide assortment of antennas to be used. The 1300H comes with rechargeable internal batteries and is very sensitive. It costs about $150 direct from Optoelectronics and is the recommended device for field work and low-profile surveillance detection. Appendix C lists other sources for frequency counters.

Warning: Using a frequency counter near certain U.S. government installations could result in serious legal problems and/or espionage charges. In foreign countries, it could result in execution.

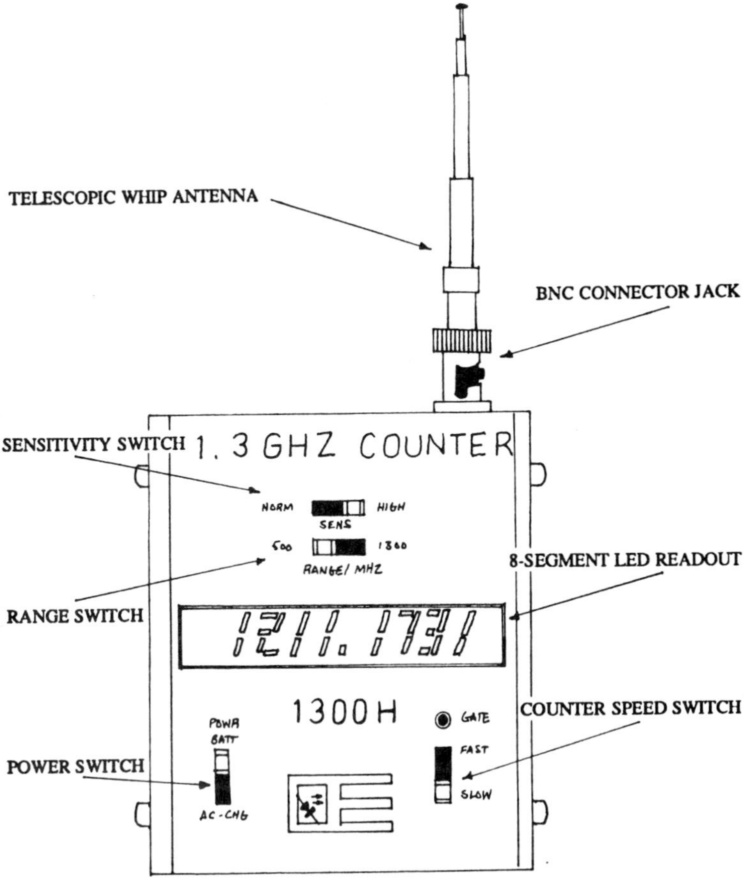

Figure 17. An Optoelectronics 1300H hand-held frequency counter (0-1300 MHz).

Chapter Nine

Police Radio Operational Procedures

Law-enforcement operations require consistent and reliable communications between field elements and headquarters and between patrol members and patrol leaders. The use of portable low-power VHF and UHF radio equipment has proven to be an effective tool in the apprehension of subjects as well as keeping patrol units accounted for on a real-time basis.

When police began using radios in 1929, the equipment was heavy and unreliable. Headquarters operated a transmitter and patrol cars had receivers. Thus, each time headquarters made a radio call, even if it was to assign a specific patrol car a specific task, it was an all-points bulletin. The patrol car had no way to confirm reception or respond. This one-way system was the only means of communication for the first ten years of police radio. Smaller police agencies had their patrol car radios tuned to local AM radio broadcast facilities and used them to contact their units in emergencies.

Modern police radio is much more sophisticated and

has become technically difficult to defeat electronically. However, individual areas have geographical limitations with regards to radio communications. There are sections of every town where police radio communications are difficult, if not impossible, to conduct. With experience in the field, patrol officers learn these areas and avoid radio use when they are there. Commonly referred to as *dead spots* or *windows*, it is important that the monitor operator make a note of and chart these areas on a map.

Radio communications are the voice and ears of a police department. Without radio equipment, the organization cannot function as a team and its combat effectiveness is greatly diminished. Police use of radio is vital to the expedient dispatch of patrol units to scenes of incidents. It allows the agency to maintain command and control over all patrol elements.

The patrol officer has a mobile radio in his vehicle and a belt-mounted walkie-talkie to use when he is on foot, making him operationally dependent upon his radio equipment. In most agencies studied, an officer will immediately get a new radio or a new vehicle if his radio is not functioning. If he is out of radio contact for *any* reason, the dispatcher will send several vehicles to his last location to investigate. Many officers also wear a new beeper device on their belt that signals an alarm if they lie down for any reason. These devices can save a life, although they also tend to expose officers slouched down in their patrol cars taking a nap.

Police radio communications are usually multichannel, but one or two frequencies are used for general communications such as license-plate number information requests, minor vehicle accident dispatches, domestic disputes, shoplifter in custody, person refusing to leave, drunk-driver reports, and animal-control problems.

These relatively minor complaints are given a priority based on severity and are generally dispatched to the closest available units. More serious police activity may have a separate, reserved channel. This might include serving warrants, inquiries for criminal records, and alarm calls. Yet another channel may be assigned to car-to-car communications during surveillance operations.

All police departments use radio slightly differently. They may stress unit accountability, or they may put out general broadcasts that are then responded to by available patrol elements. Patrol units, investigators and detectives, vice officers, traffic officers, and riot-control elements all may have specific assigned frequencies, though they may share some channels with other specialists. On the other hand, many modern operations have all frequencies available for all units. Only careful study of the target agency will uncover the specific uses of each frequency.

Police agencies know that their communications are monitored—they keep a lot of operational traffic off the radio for this reason. For instance, where they will be going to serve a warrant on an individual and when they plan to execute that warrant are *never* discussed over the radio. The same applies for raids, searches, and other operations that rely on the element of surprise.

POLICE RADIO CODES

Police radio security is based on several codes. These codes are generally specific to each agency for further security—in fact, there usually is a series of internal codes used *only* by a particular agency, and these can get more specific within different sections of that agency, such as the investigative or vice divisions. There are, however, a set of standard codes used by all police agencies, usually under the following prefix designations:

1. Ten-XX (two-digit number describing most typical traffic).
2. Signal-XXX (two- or three-digit number designating conditions).
3. Code-XX (one- or two-digit number designating condition).
4. Unit-XXX (two- or three-digit number designating individual units).

Thus, radio traffic for police officers is reduced to such jargon as "Code 4," "10-28 on a vehicle," "10-50 PI on freeway," and "Signal 101 warning." You should make careful notation of all codes and attempt to learn their meaning by the context and content of the conversation. Also note that there is often a standardization in the use of codes among local fire, rescue, and police departments. You can monitor each to learn codes that may be more common in cooperating agencies than in your target agency. After monitoring this traffic for awhile, you will acquire a good working knowledge of your target agency's radio-communication language.

Phonetic Alphabet

The Associated Public Safety Communications Officers (APCO) is a nonprofit group set up to assist law-enforcement agencies in the efficient use of radio during day-to-day operations. They have set up specific guidelines in the use of radio codes and phonetic alphabets. The following is the standard phonetic alphabet:

A	ADAM	F	FRANK
B	BOY	G	GEORGE
C	CHARLES	H	HENRY
D	DAVID	I	IDA
E	EDWARD	J	JOHN

K	KING	S	SAM
L	LINCOLN	T	TOM
M	MARY	U	UNION
N	NORA	V	VICTOR
O	OCEAN	W	WILLIAM
P	PAUL	X	X-RAY
Q	QUEEN	Y	YOUNG
R	ROBERT	Z	ZEBRA

Military Phonetic Alphabet

There appears to be a growing trend among some law-enforcement agencies to use the military phonetic alphabet. You will find that many federal agencies use these designations as well. The following is the military phonetic alphabet:

A	ALPHA	N	NOVEMBER
B	BRAVO	O	OSCAR
C	CHARLIE	P	PAPA
D	DELTA	Q	QUEBEC
E	ECHO	R	ROMEO
F	FOXTROT	S	SIERRA
G	GOLF	T	TANGO
H	HOTEL	U	UNIFORM
I	INDIA	V	VICTOR
J	JULIET	W	WHISKEY
K	KILO	X	X-RAY
L	LIMA	Y	YANKEE
M	MIKE	Z	ZULU

Police use both phonetic alphabets for reading and spelling words and names over the radio, so there should be no question as to the correct spelling or content of a message.

Ten-Codes

Ten-codes are used to keep standard messages brief and clearly understood. These codes are somewhat flexible in their use—in fact, no agency seems to use the codes exactly as they are set by APCO guidelines, although most agencies will use at least half of the standard codes. The following pages list the standard ten-codes that most agencies use or base their traffic on:

10-0	Exercise caution.
10-1	Unable to copy, change location (dead spot).
10-2	Signal good.
10-3	Stop transmitting.
10-4	OK, or acknowledgment.
10-5	Relay message.
10-6	Busy, unless urgent.
10-7	Out of service.
10-8	In service.
10-9	Repeat message.
10-10	Fight in progress.
10-11	Dog-related case.
10-12	Stop, or standby.
10-13	Weather and road conditions.
10-14	Prowler report.
10-15	Civil disturbance.
10-16	Domestic dispute.
10-17	Meet complainant.
10-18	Complete assignment quickly.
10-19	Return to station or . . .
10-20	Location.
10-21	Call on telephone (do not use radio)
10-22	Disregard last message or assignment.
10-23	Arrived at scene.
10-24	Assignment completed.
10-25	Report in person, or meet at a location.

10-26	Detaining subject, or expedite.
10-27	Drivers license information.
10-28	Vehicle registration information.
10-29	Check records for warrant or if stolen.
10-30	Unnecessary use of the radio.
10-31	Crime in progress.
10-32	Man with gun.
10-33	Emergency.
10-34	Riot.
10-35	Major crime alert.
10-36	Correct time.
10-37	Investigate suspicious vehicle or person.
10-38	Stopping suspicious vehicle.
10-39	Urgent; use lights and siren.
10-40	Silent run; use no lights or siren.
10-41	Beginning tour of duty.
10-42	Ending tour of duty.
10-43	Information.
10-44	Permission to leave for . . .
10-45	Animal carcass.
10-46	Assist motorist, or disabled vehicle.
10-47	Emergency road repair.
10-48	Traffic standard repair.
10-49	Traffic light out or malfunctioning.
10-50	Accident.
10-51	Wrecker needed.
10-52	Ambulance needed.
10-53	Road blocked.
10-54	Livestock on highway.
10-55	Intoxicated driver.
10-56	Intoxicated pedestrian.
10-57	Hit and run report.
10-58	Direct traffic.
10-59	Convoy, or escort.

10-60	Squad in vicinity.
10-61	Personnel in area.
10-62	Reply to message.
10-63	Prepare to make written copy.
10-64	Message for local delivery.
10-65	Net message assignment.
10-66	Cancellation of message.
10-67	Clear for net message.
10-68	Dispatch information.
10-69	Message received.
10-70	Fire alarm.
10-71	Advise nature of fire.
10-72	Report on progress of fire.
10-73	Smoke report.
10-74	Negative.
10-75	In contact with . . .
10-76	En route to . . .
10-77	Estimated time of arrival (ETA).
10-78	Need assistance.
10-79	Notify coroner or medical examiner.
10-80	Chase in progress.
10-81	Breathalyzer report.
10-82	Reserve lodging.
10-83	Work school crossing at . . .
10-84	If meeting, advise ETA.
10-85	Delayed due to . . .
10-86	Officer or operator on duty.
10-87	Pick up/distribute checks.
10-88	Present telephone number of . . .
10-89	Bomb threat.
10-90	Bank alarm.
10-91	Pick up prisoner or subject.
10-92	Improperly parked vehicle.
10-93	Blockade.

10-94	Drag-racing incident.
10-95	Prisoner in custody.
10-96	Mental subject.
10-97	Check signal.
10-98	Prison or jail break.
10-99	Stolen or wanted.

Other Terms and Expressions

There are certain slang or abbreviated terms that are commonly used by many police departments to describe situations or conditions. These terms have a relatively universal use:

AKA	Also known as (an alias).
ATF	The Bureau of Alcohol, Tobacco and Firearms.
B and E	Breaking and entering.
Beat	Area where officer is assigned to patrol.
Beat book	Book of detailed maps and notations regarding patrol area.
CCW	Carrying a concealed weapon.
CI	Confidential informant.
CP	Command post.
DMV	Department of Motor Vehicles.
DUI	Driving under the influence.
Dust	Fingerprint technique.
DWI	Driving while intoxicated.
EOD	Explosive Ordinance Disposal, or the bomb squad.
ETA	Estimated time of arrival.
FBI	Federal Bureau of Investigation.
Frisk	A quick search for weapons or contraband.
HP	Highway Patrol.

Hard Copy	A printout from a computer search.
HRU	Hostage Rescue Unit.
ID	Identification, or Investigative Division.
Item	Piece of evidence or other object not to be discussed on radio.
K-9	Canine unit with search or guard dogs.
Latents	Identifiable fingerprint.
Line up	Suspect identification procedure before witness or witnesses.
ME	Medical Examiner, or county coroner.
Miranda	The process of advising a suspect of his legal rights.
MO	Modus operandi; the patterns or "trademarks" of suspect or crime.
Monitor channel	Stand by on this frequency for further message.
NCIC	National Crime Information Computer.
NOL	No operators license.
OR	Own recognizance release.
Pat down	Quick search for weapons or contraband.
Package	An individual under surveillance.
Packet	Officer's paperwork for the day, including warrants, etc.
PD	Police Department.
PI	Personal injury.
Perp	Perpetrator.
Prints	Fingerprints.
Public Service	The phone company.
Radio secure	Radio to be monitored away from suspects or prisoners.

Rape kit	Hospital emergency-room kit used to collect evidence of rape.
SEU	Selective Enforcement Unit (SWAT team).
SO	Sheriff's Office.
Slim jim	A long, slender piece of steel used to open locked cars.
Stakeout	Surveillance site.
Subject	Individual.
SWAT	Special Weapons And Tactics unit.
Take down	Arrest suspect with weapons drawn or by surprise.

CODES and SIGNALS

Codes and signals generally are indigenous to an area or a department. CODE 2 and CODE 3 frequently are emergencies, CODE 4 is usually an arrest warrant, and so on. SIGNAL 44 usually is a silent alarm, SIGNAL 100 sometimes refers to a body, and SIGNAL 101 usually is a report of shots fired.

* * * * * * *

After monitoring your target agency for a period of time, you should become familiar with all of the codes in use. There are, however, a number of alternative methods and techniques that an intercept operative can use to learn the target's codes and abbreviations. Contacting scanner owners through clubs and networks can be very helpful. Many local community colleges have two-year degree programs for entry-level law-enforcement positions, and they sometimes provide a number of local codes for training. Use your imagination, and exercise a degree of care during all inquiries.

LAW-ENFORCEMENT PSYCHOLOGY

A police officer on patrol must circulate in a specific area. This geographic region may be shared with other units, or he may be on the beat alone. His primary responsibilities are to maintain a high degree of visibility; observe vehicles, pedestrians, and businesses; focus on high crime areas; and respond to calls from dispatch.

The patrol officer's daily activity is relatively mundane and incident-free. His responsibility is to stay mobile and active, even when he is not on a specific call. He spends over half his calls investigating accidents, handling civil and domestic disputes, and citing vehicles and drivers for violations. Television depictions of police work is very misleading. A typical 20-year veteran of a police department has never drawn his weapon for anything other than cleaning it and range duty. Eighty-seven percent of all police officers go through their entire careers without having to use their weapons in the line of duty.

That is not to say that police work is not dangerous. It can be stressful and frustrating, and hundreds of police officers are injured or killed in the line of duty each year. (In 1987, the largest cause of metropolitan police fatalities involving weapons was interruption of armed robberies; in smaller towns, it was interruption of domestic disputes.) After five years or so of patrol duty, however, an officer has a tendency to become somewhat complacent in his daily duties. This is extremely dangerous. Note the fact that over 20 percent of police officers who are shot in the line of duty are shot with their own weapons, according to the U.S. Department of Justice.

Police officers are acutely aware of the risks and minimal rewards of their work. They are very clannish in their private lives, generally associating only with other officers. Most are dedicated and outgoing personalities, with a deep sense of fairness and morality, yet many officers are somewhat cynical and suspicious. The psychological implications of long-term exposure to the stressful and mundane aspects of police work result in a high percentage of divorce, alcoholism, and professional burnout.

Because of the close-knit environment of most police agencies, when an officer is in any type of trouble, the agency will almost always overreact with manpower and resources. Officers tend to respond with an unusual degree of spontaneous emotion when a fellow officer is in danger or injured. They will disregard their normal assignments, even at the risk of disobeying a direct order from their patrol supervisor. This situation is even more acute among foreign police agencies. Injuring a South Korean or Chilean police officer will result in the mobilization of virtually all personnel to the scene of the incident. They will disregard all individual rights and civil liberties in their search for the parties responsible for harming one of their own.

A police officer in a foreign country basically is a soldier. His organization is usually made up of military personnel who have reentered civilian life. Individuality and personal goals are often discouraged. Although its members are heavily armed and well versed in the use and deployment of weapons, an agency would almost qualify as a soft target due to its many innate vulnerabilities. Jamming its vital communications while attacking its individual patrol elements is not technically difficult, and operations conducted by terrorist groups and revolutionary organizations against police agencies (particularly in South America) have been very successful.

TACTICAL LIMITATIONS OF JAMMING U.S. LAW-ENFORCEMENT AGENCIES

This manual does not suggest that you attack a law-enforcement agency in the United States. It is one thing to temporarily disable a segment of a department's radio communications for a brief period during an action, but it is another matter entirely to attempt to attack the agency as the primary target. American police departments have advantages that most foreign agencies do not. For instance, all agencies in the United States have what is called a Mutual Aid Agreement between bordering departments. They conduct regular emergency training exercises where one department is in "limited peril" and the other agencies respond. The domain of the assisting agencies is expanded to include the agency under siege, and their personnel are controlled by a predesignated area coordinator, such as a chief of police or county sheriff.

Many U.S. police departments also have special contingency operations and alternative means of communications, mainly due to the rioting and civil disturbances of the late 1960s and early 1970s. If an agency believes it is under siege or is the target of a conspiracy due to your UW activities, it will place all of its personnel on a serious tactical alert, and within hours your operatives will find themselves dealing with any number of interagency personnel and even federal troops.

Consultations with senior and retired law-enforcement professionals regarding radio jamming bear out the above warning. They admitted that the target agency would be severely disabled in command and control of its field elements for a short period of time. They were confident, however, that operations would eventually recover. One ranking police administrator from a major midwestern city stated that if the jamming was executed in order to commit a major crime, the odds of initial detection might indeed be seriously reduced. He added, however, that most major crimes are solved by investigation and the use of informants, and the eventual apprehension of the perpetrators would not be affected by the initial jamming.

U.S. law-enforcement agencies, therefore, primarily focus on solving crimes rather than prevention. They use sophisticated evidence-collection techniques and advanced detection and surveillance equipment, all tied together with good intelligence and computerized information systems. These disciplines and techniques are vital to modern law enforcement, but the majority of solved crimes are the result of painstaking, methodical investigation and the extensive use of informants. Informants alone result in the majority of case solutions on the municipal and federal levels in the United States.

Advanced monitoring of police radio traffic and well-planned jamming of the system has many advantages in modern guerrilla warfare, but it has severe limitations as well. Don't believe that the technical capability of jamming and monitoring is a panacea for your team. Careless deployment and poor OPSEC will result in a serious if not deadly confrontation with a well-prepared law-enforcement agency.

Chapter Ten

Police Radio Systems

Police radio systems have become very sophisticated since the mid-1980s. They are designed to provide extended range and quality as well as a high degree of electronic security.

Years ago, police agencies used the 1.7-2.0 MHz band for radio transmissions. These channels are just above the AM radio band. They slowly worked up to the FM/VHF low frequencies (30-50 MHz), where you will still find most state police, highway patrol, and rural law-enforcement radio systems. This band provides extended range compared to VHF high and UHF because the FM signals can bend over the horizon, making these frequencies ideal for highway-patrol applications. However, the antenna length required precludes the use of hand-held units and noise generated by patrol vehicles makes this an unreliable band at times. The 30-50 MHz band also is easily affected by spark-gap barrage jamming techniques.

The current trend in police radio is UHF and above for most cities and VHF high for most counties and sheriff's departments. The UHF frequencies have a much smaller wavelength, which requires a smaller antenna. They are easier to use in buildings and underground because a UHF signal can bounce off of large structures on its way to the receiver.

Police radio operations require a base station at headquarters, radios in each patrol car, and hand-held walkie-talkies for each officer and foot patrol. This system must have base-to-mobile, mobile-to-base, and mobile-to-mobile capabilities on a reliable basis. These capabilities are based on one of three systems.

The *simplex* system has all units talk and receive on the same radio frequency. Units must wait for the channel to be clear of traffic before they can talk.

Two-frequency simplex has all units transmit on one frequency and receive on another. The base frequency for receive is the mobile frequency for transmit, and vice versa.

The *repeater* system has all transmitted mobile traffic received by a large transmitter/receiver station, known as a repeater, and retransmitted on another frequency. This is commonly done on the UHF bands. The repeater frequency will usually be 5 MHz above the input frequency. The base repeater generally will be located at the target's headquarters, and a tower will be visible on the roof. This provides the agency with security for the repeater system, where if it were at another site or location, it could be targeted for attack or otherwise neutralized by criminal or enemy elements. Sometimes, however, the repeater is located at a dedicated tower on a hill or tall building instead of at headquarters. In these cases, the repeater site is connected by telephone wire to the base station.

Base repeaters are controlled by the headquarters dispatch operator. A mobile repeater is just like a base repeater except that there are no wires connected to the base. Instead, the system is controlled by radio with a directional beam antenna aimed at the repeater. The concept of a mobile repeater is the basis for our improvised jamming technique of locking up the target repeater for a specific time.

A vehicular repeater is a system that uses the patrol car's more powerful transmitter to send radio traffic from the officer's walkie-talkie back to base. Vehicular repeaters are used at temporary command posts to allow all teams involved in the operation to extend their transmit and receive range.

FREQUENCY ALLOCATIONS FOR LAW ENFORCEMENT

There are literally hundreds of published sources listing frequency allocations for most law-enforcement agencies. These frequency guides are inexpensive and available from many outlets. Appendix C contains several of these sources.

A recent FCC ruling has made it legal for law-enforcement agencies to use any frequency allocated to their assigned city or jurisdiction. When you are monitoring the frequencies, you may hear a surveillance team discussing the target of a stakeout on the city's garbage truck frequency or the county landfill frequency.

The advent of low-cost multifrequency programmable transceivers provides elements of a modern police department the ability to change their tactical frequencies randomly and quickly. Effective search/scan operations are stressed in this manual for this reason. Again, the search/scan operator should be the most experienced and knowledgeable intercept operator on the team because he will best know the typical content and even the voices in the target agency.

Many larger law-enforcement agencies also make extensive use of cellular phones in their day-to-day operations. New laws, such as the Electronic Communications Privacy Act (ECPA), have made it illegal to monitor this traffic, and agencies have begun to rely upon it more and more. This frequency band is blocked out of most scanners, and many agencies use these frequencies with a false sense of security for this reason.

The FCC has allocated specific portions of the radio spectrum for "land mobile" radio systems used by municipal, law enforcement, and business broadcasters. These bands are as follows:

1. 25-50 MHz (VHF low)
2. 150-174 MHz (VHF high)
3. 450-470 MHz (UHF)
4. 806-942 MHz (UHF/Microwave)

You will find about 98 percent of your targeted traffic in one of these relatively small frequency bands. (*Note*: 66-88 MHz is VHF low in most of Europe.)

SUBAUDIBLE TONES

All modern radio communications use a feature known as *subaudible tone squelching*. These tones are used to open the squelch when a specific tone is received. This allows the agency to share the radio frequency with different elements of the department: detectives may have one tone, vice officers may have another tone, etc. Every patrol vehicle may have a specific series of subaudible tones that identify it to the dispatcher, who sees the transmitted signal on a screen and knows which vehicle is sending the radio traffic. The tone turns on the repeater when it is received. Simply sending radio traffic on the target frequency, however, will not necessarily "kick on" the repeater to lock up the system and jam the network.

Subaudible tones can be determined by connecting an audio frequency counter to the earphone jack of your scanner receiver and noting the frequency when there is no modulation (or speech being sent), such as when an officer first keys his mike and after he is done sending traffic.

Trade names such as PL (Private Line) and Channel Guard are subaudible tone circuits that are internal to a radio or are placed into the system by the agency's radio technician. The transmitter equipment available for use in jamming is very inexpensive, and the subaudible tone can be programmed into the transmitter from the keyboard.

EIA CTCSS Subaudible Tone Frequency Chart *

FREQUENCY (HZ)	DESIGNATION
67.0	L1
69.3	WZ
71.9	L2
74.4	WA
77.0	L3
79.7	WB
82.5	L4
85.4	YA
88.5	L4A
91.5	ZZ
94.8	L5
100.0	1
103.5	1A
107.2	1B
110.9	2
114.8	2A
118.8	2B
123.0	3

FREQUENCY (HZ)	DESIGNATION
131.8	3B
136.5	4
141.3	4A
146.2	4B
151.4	5
156.7	5A
162.2	5B
167.9	6
173.8	6A
179.9	6B
186.2	7
192.8	7A

* EIA = Electronic Industries Associations
CTCSS = Continuous Tone Coded Squelch System

THE VOTER

Some police radio systems have several satellite repeaters throughout the network that provide the base with extended coverage of the patrol area of operation. Each repeater system has a device known as a *voter* that selects the strongest signal and passes it to the base station. We want our jamming signal to be strong enough so that the system "votes" for it and sends it in.

In single repeater systems, a voter may also be used for the purpose of only sending the strongest received signal back to the base unit. This is useful in eliminating a units-doubling condition where more than one unit attempts to call in at the same time, effectively jamming both transmissions.

SCRAMBLERS AND DIGITAL ENCRYPTION TECHNIQUES

Sensitive law-enforcement and military operations require secure communications. This is accomplished using various scrambling techniques.

Inversion scrambling is the oldest and perhaps simplest technique. Inversion scrambling takes the audio portion of a radio signal and puts it "out of phase," making it sound like gibberish to a listener using a regular monitor receiver. A descrambler puts the signal back "in phase" to make it readable. This technique is no longer used by police agencies because the equipment required to descramble the traffic is simple to build and easily available to the consumer. Most telephone scramblers sold in electronic catalogs are actually inversion scramblers. They sell for $200 to $500 per unit, but they are actually about $3 worth of parts inside a fancy case with lights and dials. All telephone-tapping receiver equipment used by federal and municipal agencies have circuitry that detects and then quickly descrambles this traffic. These devices are *not* secure.

Sophisticated scrambling techniques make use of digital encryption circuitry, which is almost impossible to defeat. A good example of digital encryption is the Motorola SECURENET system.

Motorola's Secure Communications Strategic Business Unit states that there are literally trillions of possible combinations that are encoded into the SECURENET system. The only way to decode the SECURENET would be to use a high-speed super computer such as the Cray computer used by the National Security Agency, and it still would take over a year of dedicated processing. Realistically, system operatives could easily enter in the twelve-digit code every day or even every shift, making it very difficult to disrupt daily communications.

SECURENET radio equipment is used extensively by covert military and law-enforcement agencies, but is also available for use by business and private enterprises. A good example is the 1988 America's Cup boat race. The U.S. crew suspected that the New Zealand team was eavesdropping on their shipboard radio traffic. Since boat speed, headings, conditions, equipment status, and other race information would have been highly useful to the competition, U.S. skipper Dennis Conner had all of his radios installed with the SECURENET system, effectively denying any party the opportunity to monitor his vital radio traffic.

To disable a scrambled system, you may think that jamming all secure traffic frequencies would be the best solution. Perhaps this might work, but a modern radio system will function even if there is other traffic on the frequency. Employing the same methods used for unscrambled (or "in the clear") traffic will not necessarily jam the system. The SECURENET, for instance, waits for secure digital traffic to occur before it allows the receiver to function. This presents a difficult technical problem in jamming.

Several communications specialists were consulted regarding this condition, and the consensus was that the jamming operator must send duplicate scrambled traffic of the format used on the target system in order to defeat the entire net. Since digital encryption occurs on the audio portion of radio traffic, the encryption instructions are a series of tones and pulses that can in fact be heard by the system operator, even if he doesn't have the ability to descramble the message.

By monitoring the scrambled traffic and recording it with a high-quality tape recording system, you can reproduce the digital traffic. You will not be able to understand its content, but you don't have to in order to jam the system—simply play back the digitally encoded signal over the target's radio frequency. The system will accept the traffic as originating from one of its transmitters in the net and attempt to decode it. This will lock up the system and prevent legitimate traffic from entering the network.

Note: It is a federal crime to monitor or attempt to decode scrambled radio traffic. Even having these frequencies keyed into a scanner is illegal!

Chapter Eleven

Air-Traffic Control and the Use of Radio

Commercial air traffic requires reliable radio communication between the air terminal and the pilot, as well as between aircraft in a specific area. The VHF AM radio band just above the normal FM band is used for this purpose. Military and commercial air traffic is also conducted in the UHF band and in the HF frequencies on upper sideband (USB).

Of all the radio systems studied, the air-traffic bands are perhaps the most vulnerable to jamming and deception. During the air-traffic controller strike in the early 1980s in the United States, there were several minor incidents of radio jamming. Although they were not directly attributed to the strike, they created a great deal of alarm among security elements of the commercial airline industry and U.S. domestic intelligence.

Radio communications equipment is found in all types of aircraft, from small twin-engine planes to jumbo jets. This equipment is easily obtainable on the open market. In most cases, if a third party operates a transmitter on a radio frequency assigned to a specific airline, authorities will notify the FCC and often will suspend operations temporarily until the transmissions either stop or the transmitter site is located.

Radar (an acronym for Radio Detection and Ranging) has many uses in modern air-traffic control. A radar control area is set up around the entire perimeter of an air terminal, where a large dish antenna revolves constantly, keeping the airspace under surveillance. Aircraft approaches, holding patterns, and stacking operations are conducted with guidance from the radar screen. If the reflector dish is damaged or fails, the air terminal will shut down immediately.

A radar beacon is a device that provides an aircraft with an automatic-pilot function. The aircraft control system locks onto the beacon and steers the aircraft through the radar corridor, providing very accurate guidance for a large aircraft. If the beacon system is out of adjustment or calibration, it will cause the aircraft to deviate from its intended route. Korean Airlines flight 007, the passenger jet that was shot down by Soviet fighters in 1983, went off course and entered Soviet airspace because its radar beacon was out of calibration. The jet wandered several degrees out of the radar corridor set up by the homing beacon.

A radar altimeter is used to determine an aircraft's altitude. This device transmits a signal to the ground and receives the same signal microseconds later. The duration of time between transmit and receive is electronically measured, and this provides an accurate reading of distance to the ground. Pilots can also control the plane to some degree by connecting the radar signals to a navigational computer. This is a basic description of automatic pilot.

Recent developments in aircraft radio jamming have focused on radar systems because most modern aircraft rely so heavily on their radar equipment. One technique that is becoming common in military applications is radar deception. Electronic signals sent by an operator cause misleading data to come across the radar operator's screen. For instance, if a signal is transmitted to the radar altimeter that is slightly slower than the normal beacon signal, the altimeter will indicate that the aircraft is somewhat higher in the air than it actually is.

Such deception can cause a serious problem. Many modern jets have warning control and automatic flap-control systems. If the aircraft breaks a preset altitude, the plane's computer automatically makes flap adjustments to compensate. This procedure is vital to managing the congested airspace over modern airports. If the aircraft adjusts too high or too low, it may cross into another aircraft's flight path.

In 1988, President Zia of Pakistan was killed when his jet suddenly crashed to the ground immediately after takeoff. Since no solid evidence of sabotage or explosives was found, ground investigators speculate that a deceptive radar signal may have been transmitted to make the aircraft's flight computer believe that the jet was gaining altitude too quickly, or that it was in the path of another jet. A false altimeter signal may have forced the computer system to adjust the flap a fraction of a degree to compensate. Since the jet was close to the ground at the time, this minor change caused the aircraft to slam into the earth.

Radar deception is beyond the scope of this manual. The sophisticated techniques and equipment required could easily fill several volumes. Should the team wish to quickly damage an airport radar installation, it can be accomplished with accurate small-arms fire directed at the radar antenna. Hitting the horn output located in the center of a radar antenna dish will easily destroy the wave guide and disable the system. The air-traffic control tower will immediately reroute all aircraft by radio, and takeoff and landing operations will be suspended.

Chapter Twelve

High-Risk Frequency Detection Techniques

The most critical area of concern for the UW team commander is that all of the target's frequency capability has been determined. In a small municipality, this may only be one or two frequencies for all agencies. In a large city, be prepared to deal with dozens, if not hundreds, of different frequencies and several bands.

Many documented cases of jamming were successful because one of the target's radio technicians unwittingly provided all of the desired frequencies. This is not likely to occur often. For practical purposes, identification of target frequencies will be accomplished using the following methods:

1. *Antenna recognition.* The size and type of antenna used by the target base and mobile units is very useful in determining their approximate frequencies and bands.
2. *Frequency counters and spectrum analyzers.* When used in proximity to the operator or base, the target's current radio frequency can be determined.
3. *Scanning.* High-speed multiband scanners can be

used to search the entire radio spectrum to locate and identify target frequencies.

If you have sufficient time to monitor and scan, the above techniques will usually provide the team with all the current frequencies used by the target net. Unfortunately, these methods often take a lot of time to carry out. It is sometimes necessary to quickly determine your target net's alternate and tactical response frequencies.

There are several deception methods that can be employed to determine this within a very short period of time. This manual will discuss only those methods that have a minimal risk to soft targets. The purpose of these deception tactics is to elicit a normal tactical response from the target as well as a response from their special emergency units. During the action phase of your operation, these seldom-used frequencies will be vital to your target. If they are effectively jammed along with normal channels, the action will probably be successful.

The following pages contain some highly controversial and extremely dangerous techniques that have been employed by guerrilla groups to determine all of their target's tactical radio frequencies. These techniques are necessary if you wish to quickly and completely understand your target's capabilities. Apply them with extreme caution.

JIGGLING THE WIRE

In intelligence jargon, when an agency or group causes something to occur to the opposition so that his response and communication equipment can be monitored, it is known as *jiggling the wire*. The better-organized terrorist units will scan, record, and log police radio traffic during their UW actions for future reference.

If your action involves the use of explosives, then the advance team must have a knowledge of EOD personnel and their frequencies. Although calling in a bomb threat to a public building is low risk and effective in getting a response, monitoring has shown that unless patrol officers and/or dogs actually locate an explosive charge or suspicious package, the bomb squad generally does not get involved. However, remote detonation of two or more small pipe bombs in an unoccupied dwelling will elicit a quick response from the entire agency, particularly if the explosion is preceded by a phoned-in threat. If your target is near a military base, the local police will probably call in the military EOD personnel.

Government agencies also use this technique. They may have phone taps and audio-surveillance devices in place at target locations, only to monitor what appear to be routine communications. Therefore, they create a diversion that will cause panic among their target personnel. They make it appear that a raid is going to occur at another location, or they actually arrest a known associate of the target. The target will frequently overreact and contact sources and associates, incriminating many parts of the organization.

The Drug Enforcement Agency once jiggled the wire to trap a New York City police detective who was providing information to a cocaine distributor. The DEA monitored the dealer's outgoing calls while sending out signals that a warrant for his arrest was imminent. He called his connection at the department for information, and this recording was used to convict the officer and the dealer.

Once a UW action has been planned and rehearsed and the day of execution has been established, you can jiggle the wire on the target at a dummy location to test and monitor his responses, and then actually jam the equipment to see how the target responds under pressure. This should be carried out close to your D-Day in order to prevent the target from taking corrective action, such as frequency changes. This will also cause slightly frayed nerves for your target, providing the action team with an exploitable advantage.

Jiggling the wire is risky and potentially compromising to the team and the operation. However, no other method is as fast or reliable to determine all of the target's operational frequencies. These deception operations can also be used in conjunction with jamming and other diversions during your action to commit large numbers of your target's personnel to out-of-the-way assignments, confuse command and control elements, and create panic and stress.

THE SNIPER

A terrorist team operating in South America is believed to have been the first to introduce this tactic. Sniper attacks in an urban area almost always produce an overreaction by local law enforcement. Conducting a countersniper operation in a heavily populated area requires virtually all available personnel and equipment. Patrol elements frequently go to riot frequencies to cordon off and secure the area. Detectives using covert surveillance frequencies participate in the operation. SWAT teams, with their tactical channels and squad-level communications, will be deployed. Emergency medical and fire frequencies, as well as their alternates, are put in use. Military and government enforcement may even be called upon.

By monitoring this critical situation by radio, it is possible to determine all of the target radio frequencies within a few hours. A mobile monitoring unit can be set up nearby to intercept even the low-power transceivers used by SWAT teams and HRUs.

Since it is an obvious security risk to use a team member as the sniper, it is often possible to simply arm and motivate an unstable person to perform this function.

Terrorist groups might use this expedient tactic, but there is an alternative technique that has been employed in Europe that we'll discuss next.

THE ROBOT SNIPER

Instead of employing an unwitting agent as the sniper, another course of action can be taken—constructing and operating a robot sniper.

Purchase an inexpensive .22 caliber rimfire rifle, such as a Ruger 10/22 or a Charter Arms AR-7. These weapons are available from gun dealers, discount chain stores, surplus and pawn shops, or from a private seller advertising in newspaper classifieds. Purchase a high capacity magazine for the rifle, such as a Mitchell 50-round clip.

From an auto parts store, purchase a 12-volt solenoid. Connect a metal rod to the armature of the solenoid and mount the assembly under the trigger guard of the rifle so that the rod pulls back the trigger when the solenoid is energized. This simple modification takes about an hour to perform. Take the modified weapon to a secluded area and test it. (There has been a problem with the feed mechanisms of most aftermarket magazines tested. This problem was eliminated by filling the Mitchell magazine with 42 rounds instead of to full capacity.)

You now have a semiautomatic rifle that can be operated by remote control. Mount the modified weapon on a secure rest—two wooden C clamps secured to a small folding table work nicely. Radio control or timers can be used to fire the weapon. (Remote control has certain advantages over timers for this application. You can observe the entire situation covertly and energize the weapon when civilians are out of the target area.)

Mount the device on a window ledge behind dark curtains. Aim the rifle at any large glass window at street

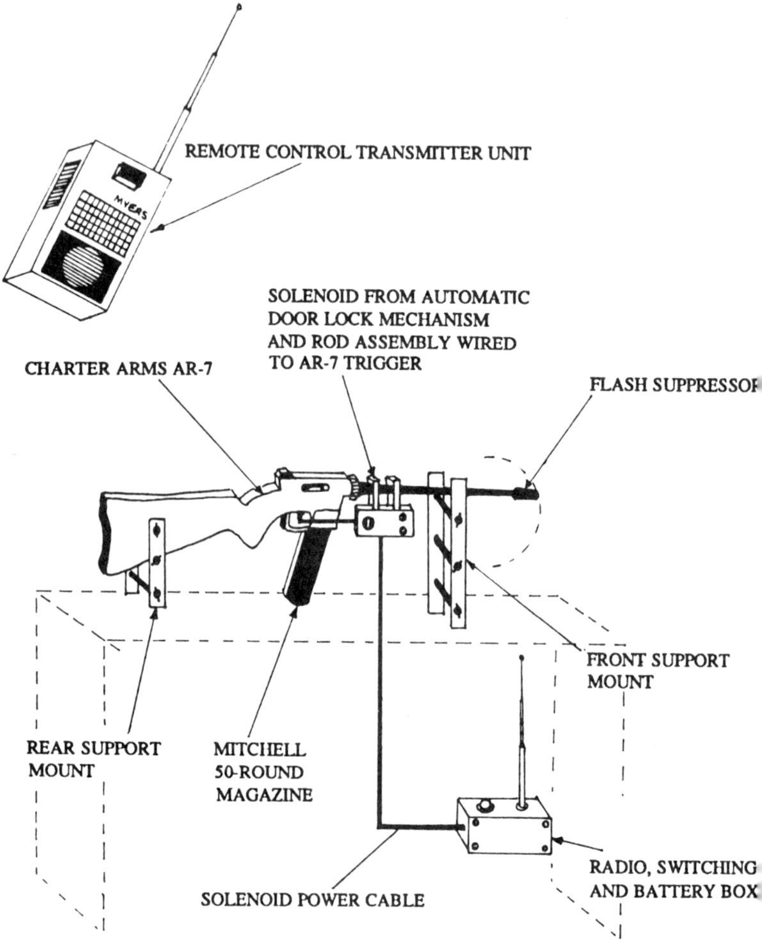

Figure 18. A robot sniper utilizing a radio-controlled .22 fifty-shot rifle. ***Warning:*** **Construction or possession of this weapon is a federal offense.**

level, clear the area, and start the incident. This action is definitely high risk, but it is an excellent way to obtain vital radio frequencies. Conducting a major action without the frequencies would be equally, if not more, dangerous to team personnel.

The robot sniper will cause havoc. If properly deployed, it will cause no injury or loss of life. If the typical response were to be documented, it would go something like this:

0000 Citizen calls 911 to report sniper or shots fired.

0002 Police dispatcher (PD) sends nearest patrol element to area. Other nearby units advise PD that they will check in with initial unit. Two to five patrol cars are immediately deployed to the area.

0007 Fire and ambulance units are called in, although there is no report of fire or injury yet.

0009 Patrol elements arrive (10-23) at scene, advise dispatch of situation, and begin car-to-car traffic with patrol supervisor on normal frequencies.

0011 Supervisor arrives. Patrol elements begin to secure area within several blocks of incident: rerouting traffic, clearing streets of pedestrians, and so on.

0013 Fire and rescue teams arrive. Supervisor advises dispatch and requests additional traffic and enforcement units.

0016 Perimeter area set up and partially cleared. Platoon officer arrives at scene, sets up a command post (CP) in building or vehicle to coordinate evacuation. Situation is assessed. SWAT personnel arrive at scene.

0018 At this point your intercept unit should be fully operational. All operators should focus on search/scan operations and logging previously unknown frequencies. The target will be using communications capability to the maximum, focusing on the following command elements:

Command Post. Area and personnel coordinator; 3-5 tactical frequencies using high-power hand-helds along with base units; 3-5 cellular mobile phone frequencies for HQ and negotiation.

SWAT Unit I. Countersniper unit; 1-3 tactical frequencies using low-power radios.

SWAT Unit II. Area-security element; 1-2 tactical frequencies using low-power radios.

Riot and Patrol Units. Area security and perimeter clearing; 1-2 tactical frequencies using normal radio equipment.

0021 SWAT countersniper units isolate specific location of sniper. CP is advised and tries to determine if there is a phone in the room to attempt contact. CP orders SWAT personnel to begin evacuation of building.

0027 Perimeter area evacuated. Traffic and crowd control progressing well. News media arrive at scene.

0031 All units are now on tactical, or secondary, channels. At least one normal frequency is secured for incident. Dispatch advises all other units to disregard incident (10-22) because manpower is adequate. Other units continue normal patrols.

0039 SWAT personnel clear building door-to-door. All floors emptied except target floor. They begin emptying target floor, starting with rooms farthest away from target room. Floor layout sent to CP.

0047 SWAT advises they have sound-monitoring equipment set up in room next door. News media begin citywide broadcast of incident.

0058 No sounds come from room. Firing has stopped.

Subject will not answer phone. SWAT personnel in adjacent building fire directly into target window while others blast hinges off of door with 12-gauge slugs. Suppressing fire stops. SWAT team sends in gas and stun grenades and enters room.

0107 Room cleared and no one is found. Weapon is located. Room is aired out with all windows and doors opened.

0110 SWAT team leader advises CP. Investigative Division (ID) is called to scene. SWAT begins search for sniper.

0123 ID determines weapon was remote fired. Radio technician called to scene to confirm device's function.

0146 Technician arrives at scene and determines remote function of weapon. CP advises dispatch to resume normal traffic on all channels. Building reoccupied by tenants.

0158 ID begins door-to-door interview of tenants and property manager. Rifle make and serial number identified over tactical channel and entered into NCIC (National Crime Information Computer) for trace.

0230 Print technicians and photographers arrive. Room vacuumed and photographed. Fibers and all other evidence sent to lab. CP advises media that sniper has escaped. No description is given, though occupant's name is given to media. CP does not advise press that weapon was remotely fired.

0239 Name and description of occupant released to all points statewide. Traffic rerouted and SWAT personnel are relieved.

Based on transcripts of radio traffic involving sniper incidents in several major cities as well as on police training documents, this sequence of events is quite accurate. If there is a telephone installed in the room, and the team is monitoring the room with a bug, it would be best to stop firing the weapon as soon as the CP attempts to make contact. The team should also monitor cellular traffic on the public service frequencies at this time, since police and other agencies have demonstrated a greater reliance on car phones during these incidents.

PHASE THREE: JAMMING

Chapter Thirteen

The Basics of Radio Jamming

Once you have determined which victim frequencies need to be jammed, it is necessary to begin setting up a covert jamming system. You will need a separate transmitter and antenna for each frequency you wish to jam continuously during your action.

Barrage jamming of the entire radio spectrum generally is not feasible due to the possibility of jamming friendly units as well as enemy units. Therefore, the best jamming is frequency-selective transmissions of a specific duration. This creates a great deal of confusion and is easily mistaken for an internal radio problem when executed properly.

Ten years ago, the high cost of land-mobile radio systems generally precluded purchasing a compatible system for jamming. Instead, guerrilla teams in the 1960s and 1970s stole enemy radios or a frequency-compatible system from a local business. This is no longer necessary, however, due to the extremely low cost of radio equipment. A jamming system capable of disrupting several channels simultaneously and frequency-selectable for quick change-over now can be purchased new or used for less than $1,000. This low cost allows the operative to set up a jamming system at a remote location, hook it up to a timer device, and clear the area. Jamming operations using this technique generally have the jamming site carefully booby-trapped with explosives, and all the equipment is sanitized of model number plates, serial numbers, and other markings.

Programmable battery-operated mobile and hand-held radio transceivers are inexpensive and easy to operate. Modified receiver antennas and homemade wire antenna systems are simple to set up and connect. They can be designed so that they are difficult to detect from air or ground observation.

The commander should select one or two operatives to function as the jamming team and a second group to function as a security element for the team. If you are dealing with a small agency in a remote area, an unattended jamming site can be set up. If you want to be able to quickly change jamming frequencies, time the jamming operation, and operate with a degree of mobility, however, you should have your command center, your monitoring station, and your jamming unit manned and capable of intercommunications.

JAMMING SIGNALS

There are several schools of thought regarding the type of jamming signal you should use, and all are noteworthy for team consideration.

Unmodulated carrier is simply a keyed-microphone transmission that sends nothing other than the subaudible tone over the airwaves. The advantage of this technique is that the victim agency will probably believe that one of its own personnel is keying the mike accidentally. Another advantage is that the target agency will frequently get on one of its covert or tactical frequencies to advise all units of the condition. This provides the team with another radio frequency to jam. Also, the target may believe that the repeater is causing the condition due to some sort of circuit failure.

The disadvantage of using unmodulated carrier is that some modern radio systems require actual voice traffic to occur before they will operate. This is known as *carrier squelch.* It is seldom used (although all new systems coming out will have it), but experimentation is required to determine if it is in use.

The next type of jamming signal is known as *noise* or *tone carrier.* When noise carrier is sent with the subaudible tone, all units hear an annoying screeching noise in their receivers, generally causing them to lock out that frequency if it continues. It is very irritating and is often attributed to internal equipment problems. The noise also stops other operators from attempting to talk over the traffic.

Harassment carrier is particularly effective because of its demoralizing effect on the target. Calling dispatch and identifying yourself as a patrol element can create instant anger once your masquerade is discovered. It also causes units to question the validity of legitimate operators when they call in, and it affects the morale of the system operators. This is risky because it is obviously a jamming condition, and the mobilization of search teams and pursuit units will occur very quickly. Playing music or recorded laughter also can be used to harass and distract the target network.

Deception is another useful jamming technique. A skilled operator can make the dispatch believe he is a patrol element, and if the dispatch station is neutralized, all patrol elements can be routed to a kill zone at an ambush site.

Another deception technique is to call dispatch and identify yourself as another police agency from a nearby county. Indicate that you were transporting a prisoner and he managed to escape after seizing your weapon, and that you are injured and in pursuit of the suspect on foot. All available units will respond.

Feedback deception is a relatively new jamming technique that has many applications. Radio traffic from the target net is recorded over several days. This recording is then played back over the airwaves, which can create a tremendous amount of initial confusion. Recording another local agency and playing back its traffic on the victim frequencies causes even more confusion, since the transmissions are attributed to either a faulty repeater or an error from the other agency.

Feedback-deception traffic of an emergency nature, such as an all-points "officer down" condition, can reroute inactive units on patrol, create general confusion, and have a demoralizing impact, since the patrol elements will recognize the voice of the dispatcher over the radio and believe it to be a legitimate radio call. Once they determine that the traffic was not legitimate, the validity of all dispatched calls comes into question, and the reliability of the radio system is compromised in the minds of patrol elements.

Repeater deception is also relatively new and can cause serious problems for the victim net. When a patrol element keys up his mike on channel one, his traffic is also broadcast on channels two, three, and four. The jamming team simply connects a scanner to the microphone of the jamming radios. Every time a patrol element calls in, its traffic is heard on all channels. This is particularly effective if you send traffic and meter-maid radio transmissions over patrol and detective channels. Again, the agency will initially believe that the problem is internal and will blame the repeater for causing all the frequencies to lock up at one time. Repeater deception using non-agency traffic, such as a local taxi company, can also be effective. The jamming is attributed to an accidental condition caused by the other agency.

Chapter Fourteen

Jamming Equipment Selection

There are dozens of radio manufacturers that make frequency-selective transceivers for use in jamming applications. We will discuss a number of inexpensive, generally available units.

There are several factors to consider when purchasing or otherwise "obtaining" your radio-jamming equipment. The primary considerations are expense, portability, durability, and whether the unit is battery powered, is frequency programmable, and has subaudible-tone capability.

Frequency compatibility is very good in the business-band transceiver market. Taxi companies, construction firms, tow-truck operations, and service companies are just a sampling of businesses that use frequencies near typical target channels. These firms generally use crystal-controlled equipment that is easily modified to disrupt the target frequencies by placing the desired crystals in the transmitter system.

Your search/scan team can help locate business transmitters operating in the desired band that would be suitable for this purpose. In South America, terrorist and guerrilla groups often hijack a few taxi cabs or business vehicles for this application. This is easily done, since most companies focus their security on merchandise, money, and men rather than their radio equipment. The radio technician in the guerrilla group then transfers the crystals to the target net and the system is ready for use.

The following table lists several inexpensive business-band radio systems that are compatible with most target frequencies. They are all portable and battery operated for covert or mobile use.

Part I: Law Enforcement/Fire/Emergency Rescue Frequencies

Model	Frequency coverage (MHz)	Output (watts)	Cost	Source
Regency RH256 BT (mobile)	150-162 MHz 16 channels	25	$300	CEI
Regency RH606 BT (mobile)	150-162 MHz	60	$430	CEI
Regency RH156 BT (mobile)	450-482 MHz 16 channels	15	$450	CEI
ICOM IC-U18 (hand-held)	450-460, 460-470 16 channels	5	$850	ICOM*
ICOM IC-U16 (hand-held)	400-430, 450-490 16 channels	3	$750	ICOM*
ICOM IC-U12 (hand-held)	450-460, 460-470 12 channels	3	$690	ICOM*

Model	Frequency coverage (MHz)	Output (watts)	Cost	Source
ICOM IC-U2 (hand-held)	450-490 2 channels	2.5	$670	ICOM*
ICOM IC-U400 (mobile)	400-430, 450-480, 480-512 16 channels	35	$870	ICOM*
ICOM IC-U200 (mobile)	450-470 2 channels	25	$700	ICOM*
ICOM IC-H18 (hand-held)	136-150, 148-165, 160-174	5	$820	ICOM*
ICOM IC-H16 (hand-held)	136-144, 148-174	3	$730	ICOM*
ICOM IC-H12 (hand-held)	150-158, 159-166 12 channels	5	$600	ICOM*

Model	Frequency coverage (MHz)	Output (watts)	Cost	Source
ICOM IC-H6 (hand-held)	150-155 6 channels	5	$590	ICOM*
ICOM IC-V200 (mobile)	148-160, 156-168, 164-174 12 channels	25	$700	ICOM*
ICOM IC-V125 (mobile)	150-158, 155-163, 166-174 5 channels	25	$610	ICOM*
ICOM IC-V100 (mobile)	136-144, 148-174, 16 channels	50	$850	ICOM*
GE Delta (mobile)	30-50, 136-174, 406-512 32 channels	110	$2,800	various dealers

Model	Frequency coverage (MHz)	Output (watts)	Cost	Source
GE Rangr (mobile)	30-50, 138-174, 406-512, 806-896 16 channels	110	$2,200	various dealers
GE MPS (hand-held)	138-174, 406-512, 64 channels	6	$2,900	various dealers
GE MPD (hand-held)	136-174, 406-512, 806-896 48 channels	5	$2,200	various dealers
GE PLS (hand-held)	30-50, 138-174, 450-470 16 channels	6	$900	various dealers
Midland Syn-Tech (mobile)	30-50, 148-174, 406-512 320 channels	110	$1,800	various dealers

Model	Frequency coverage (MHz)	Output (watts)	Cost	Source
Teletec Omni 1000 (mobile)	146-174, 440-470 512 channels	100	$2,800	various dealers
Motorola Syntor X-9000 (mobile)	30-50, 138-175, 406-512, 806-896 32 channels	110	$3,000	various dealers
Motorola Saber (hand-held)	136-174, 403-512 120 channels	6	$3,000	various dealers
Motorola MX-300S (hand-held)	138-174, 406-512 48 channels	6	$2,800	various dealers
Motorola HT-600 (hand-held)	146-174, 450-470 6 channels	5	$900	various dealers
Motorola Spectra A3 (mobile)	136-174, 450-512 99 channels	50	$1,200	various dealers

Part II: Government/Military HF Communications

Model	Frequency coverage (MHz)	Output (watts)	Cost	Source
MacKay MSR 8000	1.6-30	125	$6,000	MacKay

(*Note*: Many ham and marine radio transceivers can be easily modified for this application.)

Part III: Air Traffic Frequencies

Model	Frequency coverage (MHz)	Output (watts)	Cost	Source
ICOM IC-A20 (hand-held)	108-135 16 channels	5	$1,200	ICOM*
ICOM IC-A2	108-135 16 channels	5	$1,500	ICOM*

* ICOM sources listed in Appendix C.

CREATING A LEGITIMATE COVER

Before purchasing your radio-jamming equipment, you must create a suitable cover business that would have a legitimate use for it. Courier services, field sales, construction—any area that you already have an intimate knowledge of will be sufficient. Approach a radio equipment dealer or mail-order source (see Appendix C) with a false name and address and set up a drop point for delivery, or pick up the equipment from the dealer yourself. If possible, use a series of "cutouts" for this purchase so that the ultimate destination is difficult to trace.

Several radio dealers were approached for equipment purchases during the development of this manual, and they all were glad to have the business. None requested any sort of identification or business license before they would sell the equipment. Advise the dealer of your needs and ask about used radio equipment that is keyboard programmable. All the dealers surveyed had many such units available, although they all wanted to sell new models. They also will automatically file for and obtain the necessary licensing documents from the FCC for you.

The model selected for use in your operation may have an internal circuit that prohibits the user from changing the frequencies. This is generally a diode on the main circuit board that must be clipped. Either secure a complete service manual (which is always highly recommended) or have the radio shop remove this diode before the gear is delivered. Many legitimate companies need a frequency-selectable unit, such as a contractor who deals with many different construction firms and needs access to their various radio systems. None of the dealers interviewed had any problem with making the units programmable upon request, although most charged $10 to $50 for this simple, quick service.

After the dealer has personally demonstrated the unit's frequency programming and subaudible tone selection, pay him in cash and provide him with a mail drop for the FCC papers, factory warranty data, and other documentation.

EQUIPMENT REQUISITIONING: A CREATIVE APPROACH

There are several creative methods to secure the necessary radio transmission equipment for a jamming operation. Illegal means are obviously risky and potentially compromising to the operation, but they do provide a level of OPSEC, just as stealing a vehicle at random provides disposable transportation for an operation.

One less-expensive route to take when acquiring radio equipment is to explain your needs to a dealer over the phone and convince him of the large purchase you intend to make. Ask him for the names of companies that have purchased similar equipment as references to his good service and business practices. Most dealers have ongoing relationships with several local firms, and it is a fairly standard practice to provide references. In fact, it is probably a source of pride to the dealer, since two-way radio sales are very competitive.

Once you've obtained a list of several local firms that have frequency-programmable radio systems, simply select one or two that have low security and steal a couple of units out of their vehicles. This is risky, of course, but it provides an excellent cutout, since there will be no record of you purchasing the equipment. You needn't steal the power cables or antennas—just steal the main system mounted under the dash (or in the trunk or backseat of the vehicle if it is an older set).

If the mission is a one-time, quick operation, another route can be considered. Guerrilla groups have been known to forcefully take over small businesses that operate high-powered, two-way radio base stations within range of the target agency. They then change the transmit frequencies to jam the target, lock down the radios on these frequencies, and clear the area.

Chapter Fifteen

Jamming Equipment Deployment

WARNING: Possession of any of the transmission equipment described in this section is a criminal act. Use of any transmitter device to disable emergency communications will result in vigorous pursuit and prosecution.

Programming a transceiver is as easy as programming a scanner. Follow the instructions in the owner's manual for the specific unit you have selected. At least three different controls have to be programmed into the keyboard: frequency (in MHz), subaudible tone (in Hz), and power output (high or low).

Learn to program the transceiver for a specific set of frequencies and subaudible tones. This procedure should be fast and error free. Once you are convinced of your ability to program the transceiver, clear it and begin the sanitizing process.

SANITIZING YOUR EQUIPMENT

Due to the inherent risks involved with jamming, you must remove all serial number and model number plates from the case and internal circuitry on the radio system to further thwart any investigator who happens to locate your jamming site. All radio transmitters have a metal plate attached with steel grommets to the back or underside that describes the device, its coverage, output power, and manufacturer's FCC identifier. Simply file the tops off each grommet and pry the plate off.

Inside the radio is another set of serial and model numbers, either printed on a small paper label or embossed directly on the main circuit board. Use a knife to remove the label or completely scratch off the embossed information.

It is not necessary to remove the model number and manufacturer's name from the faceplate on the front panel, since the radio's internal circuitry and layout configuration are easily identifiable to a technician. You only need to conceal its point of purchase or theft. Sanitize your intercept equipment in the same manner, and don't overlook the computer system and other peripheral gear.

EQUIPMENT TESTING

After you are confident in your ability to operate and program your equipment, it is time to conduct a series of tests with the system. Select either a discone or beam antenna and connect the system using RG-6U coax, keeping your feedline as short as possible.

Have your search/scan team provide you with an unused frequency in your area of operation. Program the transceiver for this frequency and key the mike while your monitor team listens. Keep this transmission *very short*, since it is riskier to transmit on an unassigned frequency than it is on a normal business channel. Verify that the transmission can be copied some distance from the monitor site. Use your wattmeter to check SWR (standing wave ratio) and your frequency counter to verify output frequency. Coordinate this operation using your interstation radio equipment.

Once you have a fully operational system, it is time to conduct your second series of tests. Have your monitor team assist you in locating a soft civilian target to practice on, such as a taxi company or delivery service. Many suitable practice targets will use city-wide repeater systems and subaudible tones.

Program in the target frequencies and monitor the base-to-car traffic for a while in order to understand the various call signs and codes. When you are comfortable with their traffic, attempt to call one of the mobile units by number. If the channel is shared with other services, the target probably will be using subaudible tones. Use your audio-frequency counter to determine this tone. (*Note*: While most modern transceivers have internal filtering circuitry that eliminates this tone before it gets to the speaker or earphone jack, most scanners do not. Use your monitor equipment for this application.)

If you detect a subaudible tone, program it into your transceiver and attempt to contact one of the base or mobile units by radio. They generally respond quickly. Once you get a verbal response from one of the operators, do not call again. Don't even acknowledge the target's ability to hear your radio transmission. At this point, you should move your jamming site to another secure location.

It is unlikely that the practice target will make a notation of your transmission if it does not reoccur. The purpose of this test was simply to verify your ability to break the squelch through the subaudible tone system of a random target. Once you are redeployed at another temporary location, it is time to conduct your third operational test.

Locate a low-level government repeater system in your

target area, such as the municipal garbage truck or water-service radio system. The frequencies for these services are available in most frequency guides (see Appendix A) or can be found simply by monitoring the bands.

Once you have located the target system, reprogram your unit to transmit out on this frequency with the proper subaudible tone. Incidentally, if you have problems determining your target system's subaudible tone, then try them all. There are only thirty or so possible tones that can be used—just keep sending them out until you kick on the repeater. It is a tedious process, but it is sometimes necessary for field applications.

After you have everything set up and tested, send an open mike (unmodulated carrier) over the repeater-input frequency. Have your monitor team verify that the repeater is working with your signal. If the repeater does not kick on, then briefly whistle into the mike to see if the system has a carrier-squelch circuit on line. If the repeater kicks on after you whistle, then it is probable that other systems in your target city will have carrier-squelch circuitry as well.

FINAL DRY RUN ON THE TARGET

You now can quickly program and clear your system, and you have a good antenna and a working transmitter. You know that the subaudible-tone transmission is functional, and you have the ability to lock up a repeater close to your ultimate target frequency. Now comes the final, risky test.

Select a time when your target's radio traffic is at an extreme low. For law enforcement, this is typically between the hours of four or five in the morning. Program in your target frequencies and direct your beam antenna toward the target repeater. Key the mike and have your monitor team listen on the repeater's output signal. *Do this only for a couple of seconds.* If the system functions well, then program in another target frequency and key up again. Once you are convinced that you have the ability to jam all of the target agency's frequencies, then take your site down, sanitize the area, and leave. Never return to this area again.

It is possible that the target agency will assume that these brief transmissions were accidental mike keys by one of their own patrol elements. It is also possible, however, that a government monitor recognized your transmission as coming from a fixed directional transmitter, and was able to get a quick triangulation fix on your position. Therefore, it is imperative that you conduct your testing as quickly as you can and clear the area.

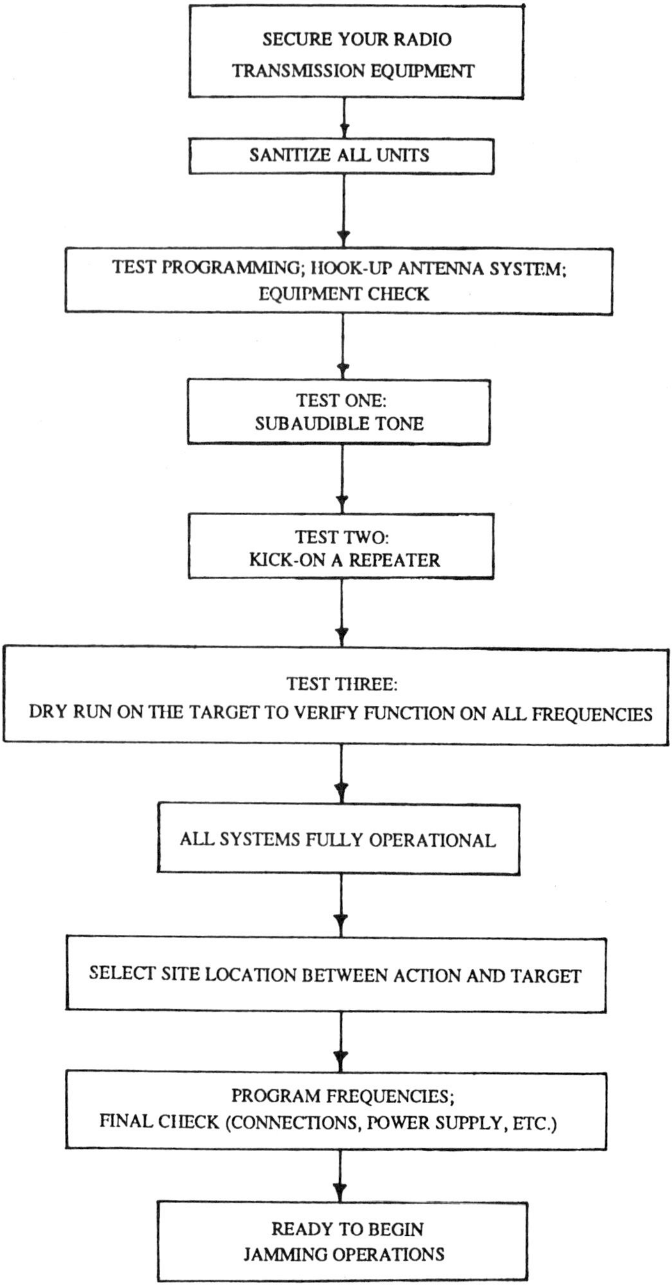
SECURE YOUR RADIO TRANSMISSION EQUIPMENT
SANITIZE ALL UNITS
TEST PROGRAMMING; HOOK-UP ANTENNA SYSTEM; EQUIPMENT CHECK
TEST ONE: SUBAUDIBLE TONE
TEST TWO: KICK-ON A REPEATER
TEST THREE: DRY RUN ON THE TARGET TO VERIFY FUNCTION ON ALL FREQUENCIES
ALL SYSTEMS FULLY OPERATIONAL
SELECT SITE LOCATION BETWEEN ACTION AND TARGET
PROGRAM FREQUENCIES; FINAL CHECK (CONNECTIONS, POWER SUPPLY, ETC.)
READY TO BEGIN JAMMING OPERATIONS

Chapter Sixteen

Jamming Operations

After you have determined that the system will, in fact, jam the target, it is time to select your covert jamming site. Select your site based on security and elevation, ideally somewhere near the midpoint between the target and the planned action. Jamming from this location will eliminate repeater, car-to-car, and car-to-base traffic, particularly traffic close to your operational area. Set up your site at least twenty-four hours prior to the action so that your jamming team can hook up and test the system before putting it into operation. Keep in mind that the jamming will be most useful during and immediately after the action.

Once the action has been completed, the jamming team should remove as much equipment as possible and sanitize the area. Destroying the site will attract too much attention. Unless the area is clearly compromised, it is not advised.

OPERATIONAL TECHNIQUES

The jamming process is simple. Identify and intercept the target's radio traffic. Analyze his communications system and use of radio. Determine his tactical requirements and the limitations of his radio equipment. Obtain frequency-compatible equipment and, at the right moment, defeat his system. *Identify, Intercept, Analyze,* and *Defeat* are the basic concepts of tactical jamming.

Keep your operational planning and intentions very secure. The greatest flaw in all jamming operations studied was poor OPSEC. This cannot be stressed enough.

Remote-Control Jamming

There is a way to operate an unattended radio-jamming site that does not have the restrictions that are inherent with the use of timers. Simply connect the beeper section of an automobile-alarm pager device directly to a 5V DC relay and connect the transmitter power cables up to the relay. When the alarm is energized, the jamming system will function until it is located and dealt with by the target.

This is a sophisticated technique, but if security or personnel is limited, it is a relatively easy route to take. Booby-trapping the jamming site is required.

Improvised Radio Detonation Techniques (available from Paladin Press) covers several remote-control applications that may be useful for the UW operative.

Mobile Jamming Operations

According to standard operating procedure, a police patrol officer is required to advise dispatch of his location and provide a description and tag number of a vehicle he wishes to pull over before he is permitted to get out of his car. This protects the officer should he encounter a violent or armed individual. This can change from area to area. Some jurisdictions require the officer to radio in before he turns on his lights or siren, and some require him to radio his position and intentions after he has the vehicle stopped.

A common practice is for a police officer to call in the vehicle license-plate numbers before even considering a stop, just to see who he might be pulling over.

Prior knowledge of a random patrol officer's operating frequency is difficult to determine when traveling from state to state. Yet there is a means of detecting and identifying the frequency of a police radio transmission just by being near the vehicle. By connecting frequency counter's input up to your car radio aerial antenna, you will get a quick readout of all transmissions originating within a couple hundred feet of your vehicle. This procedure, in conjunction with a good radar detector, is an excellent early warning system if you are conducting a long-range hostile extraction or carrying compromising materials or personnel over the open road.

In practice, using a portable frequency counter while driving is somewhat difficult. If you encounter suspected law-enforcement radio traffic, it is virtually impossible to quickly program the jamming transceiver for the frequency while keeping your eyes on the road. What is needed is a degree of planning by utilizing two vehicles for any high-risk movements. A mobile monitoring and jamming unit should always be in sight in front of the contraband-carrying vehicle.

Ideally, the jamming unit should be a windowless van, motor home, or covered pickup truck. A fiberglass-capped pickup is probably the best vehicle, since your monitoring and jamming antenna system can be concealed below the roof. All equipment should be carefully concealed, yet accessible for quick use. A two-man jamming team should stay in the back of the vehicle, and a window to the cab section is recommended. The driver maintains radio contact with the team vehicle, at all times keeping it in sight in the rearview mirror.

From their vehicle, the jamming team conducts intercept and early-warning operations. Using frequency guides and search/scan operations, the monitor operator programs in all known radio frequencies of each law-enforcement jurisdiction before the convoy enters into that area. A high-capacity scanner is ideal for this application.

The second operator monitors all likely frequencies on a multifrequency transceiver, monitors frequency-counter readout, and communicates with the driver. Should the local police inquire about the team vehicle, it will be heard by the monitor and detected by the counter operator. The jamming transceiver is keyed for no more than ten seconds after the officer makes the inquiry, and then it is shut down. Since the officer cannot hear the reply from dispatch, he has no way of knowing if his traffic was received. As long as the team vehicle is not in clear violation or suspected of felony activity, the officer will simply disregard the call and continue on with his patrol, or get off at the next exit and attempt to call in again (officers frequently encounter a temporary loss of radio communications due to location, weather, etc.). He may even stop by the nearest phone and advise dispatch that his radio is not functioning.

During a hostile extraction from any type of high-risk action, it is ideal to have mobile jamming capability in conjunction with a base jamming site directed at the target agency's repeater. You then have the ability to disable communications by pursuit vehicles conducting car-to-car radio traffic, as well as denying them the ability to radio their location and intentions in to HQ.

Good planning is essential for this type of operation. Once the target determines that you are jamming its radio traffic, the operation becomes much riskier. Jamming without the element of surprise is substantially less effective.

Broadcast-Radio Jamming

Jamming a radio station requires several wireless transmitters tuned to the target frequency and placed throughout the area. When the system is operational, the transmitters will cause the station's broadcasts to sound like a loud tone. This condition, known as *heterodyning*, will quickly cause the listener to tune in another station or turn off his radio.

There are many manufacturers of wireless microphone devices in the United States and abroad. They are inexpensive and designed to operate in the same frequency range as AM and FM radio broadcasts. Appendix C lists several sources for wireless microphones, as well as low-power radio station broadcast equipment that can be used for legal or clandestine broadcasting purposes.

Certain large urban radio stations have been known to use the above technique to cheaply and effectively lower the Arbitron ratings of their competition. This is illegal, but almost impossible to prove or prevent. The U.S. government does the same thing to unfriendly radio stations abroad. After jamming their signal, they broadcast a similar-format program close to the target frequency, a covert technique known as *piggyback broadcasting*. The listening audience believes that they simply need to retune their receivers for the desired station, not realizing that they are listening to another station entirely.

Air-Traffic Control Radio Jamming

A low-power, hand-held radio can be used to harass frequencies used by major commercial airlines for flight control, ground control, plane-to-plane transmitting, and other communications. Any disruption of communications between aircraft and the control tower will cause a great deal of alarm at the targeted airport. The Federal Aviation

Administration (FAA) and their foreign counterparts consider this to be a prime area for potential terrorist operations.

Feedback deception using actual control-tower traffic can create havoc. One common method is to jam the control tower radio frequency and send deceptive data over a nearby frequency. False weather and telemetry information can be transmitted using this technique.

Because of the sensitive nature of aircraft communications, it doesn't take a long period of false radio traffic before the terminal suspends its operations. Most pilots will refuse to fly if they believe there is a possibility of receiving false signals in-flight. Thus, shutting down a major airport can be done easily and cheaply, with no serious injury or loss of life. Appendix B lists most of the standard radio frequencies used by major commercial airlines.

OPERATIONAL LIMITATIONS OF RADIO JAMMING

You have your Mission Warning Order formulated, which indicates the time of departure, route, rules of engagement, military objectives, and other operational details. Your planning includes intermittent jamming of the agencies you have determined will be patrolling your target area. You have set up several diversions and manpower-straining distractions that will keep many patrol elements occupied away from your target area. You have monitored and studied your jamming victim for several weeks and have created an accurate assessment of its manpower and typical response time.

Your covert monitor site is manned and ready to monitor radio traffic that manages to get through during the action. Your jamming site is tested and fully operational.

Security elements at each location are on full alert.

Your action team has rehearsed every aspect of the mission so that it flows like clockwork. They are prepared to deal with the contingency of jamming failure or a hostile extraction based on a stray patrol element observing the action. You are completely confident in your planning and personnel.

Before giving the green light, there is something you should carefully consider. Although jamming the target will cause him to be less capable of coordinating an effort toward you or your team, he will not be completely disabled, just as you would not be completely disabled if your radio communications were jammed. Keep this in mind. Don't rely too heavily on your jamming capability or you will defeat the true purpose of the operation.

Any combat veteran will tell you that when radio equipment is being jammed, it is an almost certain indication that the enemy is about to conduct a hostile operation near the target of the jamming. Law-enforcement officers recognize this as well, and this concept is an integral part of any agency's training doctrine. Jamming may seriously disable the target's command and control capabilities, but it also indicates a need to be on a high state of alert.

Law-enforcement agencies generally have a contingency plan in case of radio jamming or a system-wide radio failure. Patrol elements are automatically sent to strategic locations, and critical areas are secured. Alternate radio equipment and frequencies (or some other means of communication) are brought into operation.

One experienced Special Forces operative compared a radio jamming unit to a sniper. The opposition's first objective is to find him—and, like a sniper firing his weapon, every time you key up the jamming transmitter, you are giving away your position. This giveaway is most acute during mobile jamming operations, because your jamming equipment acts like a homing device that the opposition can use to track your movements and locate your position.

It would be a fatal mistake, therefore, to rely too heavily on your ability to intercept and jam enemy communications. For every documented case of radio jamming that produced results in securing a military objective, there are at least two similar cases where jamming contributed to the demise of the operators due to inadequate or amateurish OPSEC. Overemphasis on complex technical capabilities has become almost a cliché in explaining mission failures in the after-action reports of special operations.

* * * * * * *

Under the right conditions, a well-trained jamming team operating in a concealed location with the element of surprise on its side can do incredible damage to an unsuspecting target. The success of any jamming mission is based on flawless OPSEC, excellent intelligence, careful planning, and well-rehearsed execution. Aggressiveness, innovation, technical superiority, and the element of surprise are vital factors for the mission's success.

APPENDICES AND INDEX

Appendix A

Publications

FREQUENCY GUIDES (updated annually)

Air Scan
Tom Kneitel
(CRB Research) *
120 pages

Guide to Embassy and Espionage Communications
Tom Kneitel
(CRB Research)
96 pages

Guide to Utility Stations
J. Klingenfus
(Universal Shortwave)
482 pages

Passport to World Band Radio
Radio Data Base International
Larry Magne and Tony Jones
(Grove Enterprises)
400 pages

Police Call Radio Guide
Hollins Radio Data
(Radio Shack)
9 volumes covering all 50 states

The Secret Shortwave Spectrum
Harry Helms
(Universal Shortwave)
243 pages

Shortwave Directory
Bob Grove
(Grove Enterprises)
200 pages

Top Secret Registry of U.S. Government Frequencies
Tom Kneitel
(CRB Research)
192 pages

U.S. Military Radio Communications
Michael Schaay
(Universal Shortwave)
3 volumes

World Press Frequencies
Thomas Harrington
(Universal Shortwave)
82 pages

(* See Appendix C for all source addresses)

MILITARY AND TECHNICAL PUBLICATIONS

Basic Communications Principles
SS0450
U.S. Army Signal Corps
Fort Gordon, Georgia

Combat Intelligence
FM 30-5
U.S. Army

Electronic Warfare Handbook
ST 7-181
U.S. Army

Field Radio Techniques
FM 24-18
U.S. Army Signal Corps
Fort Gordon, Georgia

Radio Wave Propagation
SS325
U.S. Army Signal Corps
Fort Gordon, Georgia

Small Unit Leaders Counterinsurgency Handbook
NVMAC 2641
U.S. Marine Corps Institute

Tactical Communications Data
CRC-500
U.S. Army Field Artillery School
Fort Sill, Oklahoma

Tactical Communications Doctrine
FM 24-1
U.S. Army Signal Corps
Fort Gordon, Georgia

Tactical Radio Systems
SS014
U.S. Army Signal Corps
Fort Gordon, Georgia

GENERAL PUBLICATIONS

ARRL Handbook
American Radio Relay League

The Codebreakers
David Kahn
Macmillan Publishing Company

Electronics Data Book
Tandy Corporation
Radio Shack

Landmobile and Marine Technical Manual
Ed Noll, W3FQJ
Tab Books

Police Patrol: Tactics and Techniques
Thomas F. Adams
Prentice Hall

Radio Handbook
Bill Orr, W6SA1

MAGAZINES

Monitoring Times
P.O. Box 98
Brasstown, NC 28902

Popular Communications
76 North Broadway
Hicksville, NY 11801

Appendix B

Classified Government and Commercial Radio Frequencies

Monitor clubs and other groups have found literally thousands of classified government and commercial radio frequencies through search/scan operations. Another way to learn these frequencies is to scan any that have been given a "classified use" designation by the FCC (found in most frequency guides). The FCC will not provide the names of the users of most of these frequencies.

Warning: The Electronic Communications Privacy Act (ECPA) states that monitoring some of the following radio frequencies is a federal crime.

FORMAT:
AGENCY NAME
Search/Scan frequency ranges for locating other channels
Documented channels for HF, FM, VHF, and UHF

UNITED STATES AIR FORCE

Search/Scan ops 225.00-400.00 MHz

4725 kHz USB	Strategic Air Command (refueling operations)
4517 kHz USB	Military Affiliate Radio Service (MARS)
4593 kHz USB	MARS
4703 kHz USB	Tactical Air Command
6761 kHz USB	Strategic Air Command (primary night frequency)
6731 kHz LSB	Air Force I (President's plane)
6756 kHz LSB	Air Force I (President's plane)
8967 kHz LSB	Air Force I (President's plane)
9018 kHz LSB	Air Force I (President's plane)
11180 kHz LSB	Air Force I (President's plane)
13201 kHz LSB	Air Force I (President's plane)
13215 kHz LSB	Air Force I (President's plane)
13247 kHz LSB	Air Force I (President's plane)
15048 kHz LSB	Air Force I (President's plane)
18027 kHz LSB	Air Force I (President's plane)
18390 kHz USB	Air Force II (VIP transport)
11055 kHz USB	Air Force II

	(Secretary of State)
6818 kHz USB	Air Force II (VIP transport)
3116 kHz USB	Air Force II (VIP transport)
13247 kHz USB	Air Force II (VIP transport)
6918 kHz USB	Air Force II (VIP transport)
9027 kHz USB	Strategic Air Command
11215 kHz USB	AWACS Radar Aircraft
11243 kHz USB	Strategic Air Command (primary daytime frequency)
13241 kHz USB	Strategic Air Command
15041 kHz USB	Strategic Air Command
17075 kHz USB	Strategic Air Command
20631 kHz USB	Strategic Air Command
6683 kHz USB	Andrews Air Force Base (routine traffic)
6927 kHz USB	Andrews Air Force Base (routine traffic)
46.75 MHz VHF	President's helicopter
122.75 MHz VHF	Air-to-air traffic
122.90 MHz VHF	Government aircraft
143.46 MHz VHF	MARS
236.60 MHz VHF	Air Force control towers
364.20 MHz UHF	NORAD (primary channel)
264.90 MHz VHF	NORAD (secondary channel)
266.50 MHz VHF	Strategic Air Command (refueling)
272.70 MHz VHF	Flight weather

257.80 MHz VHF	Control tower calling
311.00 MHz UHF	Strategic Air Command (primary)
321.00 MHz UHF	Strategic Air Command (secondary)
415.70 MHz UHF	Air Force I to ground stations

UNITED STATES SECRET SERVICE: DEPARTMENT OF THE TREASURY

Search/Scan ops 162.00-172.00 MHz

162.8500 MHz	White House security detail
167.8250 MHz	White House security detail
164.8850 MHz	President's limousine channel oscar
162.6850 MHz	Air Force I detail
171.2350 MHz	Air Force I detail
162.3750 MHz	Hand-held
162.6850 MHz	Hand-held
163.3600 MHz	Hand-held
163.4000 MHz	Hand-held
163.8100 MHz	Hand-held
164.7500 MHz	Hand-held
164.8850 MHz	Hand-held
165.0250 MHz	Hand-held
165.0850 MHz	Hand-held
165.2100 MHz	Undetermined
165.2350 MHz	Undetermined
165.3750 MHz	Undetermined
165.6750 MHz	Hand-held
165.6850 MHz	Hand-held
165.7600 MHz	Mobile
165.7850 MHz	Mobile

165.9000 MHz	Undetermined
165.2100 MHz	Undetermined
166.2100 MHz	Mobile
166.4050 MHz	Mobile
166.5100 MHz	Mobile
166.6150 MHz	Hand-held
166.7000 MHz	Hand-held
168.4500 MHz	Mobile
169.6250 MHz	Mobile
171.2350 MHz	Mobile

UNITED STATES DRUG ENFORCEMENT ADMINISTRATION (DEA)

Search/Scan ops 163.00-166.00/171.00-173.00/ 415.00-420.00 MHz

7657 kHz USB	Foxtrot (night frequency)
11076 kHz USB	Echo
14686 kHz USB	Papa (day frequency)
18666 kHz USB	Hotel (long-range day frequency)
23402 kHz USB	Romeo
163.185 MHz	
163.535 MHz	
165.235 MHz	
165.285 MHz	
165.290 MHz	
172.000 MHz	
172.005 MHz	
172.200 MHz	
415.600 MHz	

416.050 MHz
416.200 MHz
418.625 MHz
418.675 MHz
418.700 MHz
418.725 MHz
418.750 MHz
418.775 MHz
418.800 MHz
418.825 MHz
418.875 MHz
418.900 MHz
418.925 MHz
418.950 MHz
418.975 MHz
419.000 MHz

UNITED STATES CUSTOMS SERVICE ANTI-SMUGGLER OPERATIONS

Search/Scan ops 165.00-167.00 MHz

5571 kHz USB	Yankee Bravo (short-range daytime)
8912 kHz USB	Yankee Charlie (daytime backup)
11288 kHz USB	Yankee Delta (used for tracking aircraft and ships)

165.3275 MHz
165.2875 MHz
165.5375 MHz
165.7375 MHz
166.4375 MHz

166.4625 MHz
166.5875 MHz

FEDERAL BUREAU OF INVESTIGATION (FBI)

Search/Scan ops 148.00-169.00/405.00-420.00/
467.00-469.00 MHz

7905 kHz USB	167.220 MHz
9240 kHz USB	167.235 MHz
10500 kHz USB	167.250 MHz
149.375 MHz	167.260 MHz
163.310 MHz	167.275 MHz
163.485 MHz	167.285 MHz
163.810 MHz	167.300 MHz
163.825 MHz	167.310 MHz
163.835 MHz	167.325 MHz
163.875 MHz	167.335 MHz
163.885 MHz	167.360 MHz
163.910 MHz	167.375 MHz
163.925 MHz	167.385 MHz
163.935 MHz	167.395 MHz
163.950 MHz	167.400 MHz
163.960 MHz	167.410 MHz
163.975 MHz	167.425 MHz
163.985 MHz	167.435 MHz
164.260 MHz	167.450 MHz
164.275 MHz	167.460 MHz
164.410 MHz	167.475 MHz
164.460 MHz	167.485 MHz
165.525 MHz	167.500 MHz
166.500 MHz	167.510 MHz
167.150 MHz	167.525 MHz
167.210 MHz	167.550 MHz

167.560 MHz
167.575 MHz
167.585 MHz
167.600 MHz
167.610 MHz
167.625 MHz
167.635 MHz
167.650 MHz
167.660 MHz
167.675 MHz
167.685 MHz
167.700 MHz
167.710 MHz
167.725 MHz
167.735 MHz
167.750 MHz
167.760 MHz
167.775 MHz
167.785 MHz
167.805 MHz
167.875 MHz
167.925 MHz
167.985 MHz
168.885 MHz
406.200 MHz
406.250 MHz
406.275 MHz
406.300 MHz
406.325 MHz
406.350 MHz
406.375 MHz
406.400 MHz
406.450 MHz
408.850 MHz
408.875 MHz
408.900 MHz
408.925 MHz
408.950 MHz
408.975 MHz
409.000 MHz
409.100 MHz
409.150 MHz
409.175 MHz
409.200 MHz
409.250 MHz
411.025 MHz
411.075 MHz
412.425 MHz
412.450 MHz
412.475 MHz
412.500 MHz
412.550 MHz
412.575 MHz
413.425 MHz
413.550 MHz
413.975 MHz
414.000 MHz
414.025 MHz
414.050 MHz
414.075 MHz
414.100 MHz
414.125 MHz
414.150 MHz
414.175 MHz
414.200 MHz
414.225 MHz
414.250 MHz
414.275 MHz

414.300 MHz	417.550 MHz
414.350 MHz	419.200 MHz
414.375 MHz	419.225 MHz
414.400 MHz	419.250 MHz
414.425 MHz	419.275 MHz
414.450 MHz	419.300 MHz
414.475 MHz	419.325 MHz
414.500 MHz	419.350 MHz
414.525 MHz	419.375 MHz
414.575 MHz	419.400 MHz
415.750 MHz	419.425 MHz
417.075 MHz	419.450 MHz
417.150 MHz	419.475 MHz
417.400 MHz	419.500 MHz
417.450 MHz	419.575 MHz
417.500 MHz	467.950 MHz

CENTRAL INTELLIGENCE AGENCY (CIA)

Search/Scan ops	163.00-166.00/ 407.00-409.00 MHz

163.810 MHz
165.010 MHz
165.110 MHz
165.385 MHz
408.600 MHz

ISRAELI INTELLIGENCE (MOSSAD)

13920 kHz
7446 kHz
10124 kHz

ARMORED CAR FREQUENCIES

159.496 MHz	Brinks Trucks
159.600 MHz	Wells Fargo
460.975 MHz	Wells Fargo

BUREAU OF ALCOHOL, TOBACCO AND FIREARMS (BATF)

165.2875 MHz

U.S. MARSHAL

163.200 MHz

INTERNAL REVENUE SERVICE (IRS)

165.950 MHz

FEDERAL PRISONS

170.875 MHz
170.925 MHz

COMMERCIAL AIRLINES

Search/Scan ops 108.00-136.00 MHz

129.200 MHz	American Airlines
129.300 MHz	United Airlines
129.500 MHz	United Airlines

129.550 MHz	Delta Airlines
130.100 MHz	Delta Airlines
130.900 MHz	Continental Airlines
130.950 MHz	Eastern Airlines
131.450 MHz	Delta Airlines
131.925 MHz	Federal Express

GOODYEAR BLIMP

132.000 MHz	
161.640 MHz	
469.9125 MHz	(security)

PIRATE RADIO FREQUENCIES

26995 kHz
27045 kHz
27095 kHz
27195 kHz
49.875 MHz
151.625 MHz
151.6425 MHz
464.500 MHz
464.550 MHz
444.000 MHz
156.875 MHz
156.900 MHz

Appendix C

Equipment Sources

Ace Communications
10707 East 106th Street
Indianapolis, IN 46256
(800) 445-7717
(317) 849-2570

American Radio Relay League
225 Main
Newington, CT 06111

Communications Electronics, Inc.
P.O. Box 1045
Ann Arbor, MI 48106
(313) 973-8888

CRB Research
P.O. Box 56
Commack, NY 11725

Electron Processing, Inc.
P.O. Box 708
Medford, NY 11763
(516) 764-9798

Electronic Center
Ross at Central Expressway
Dallas, TX 75201
(214) 969-1936

Electronic Equipment Bank
516 Mill Street NE
Vienna, VA 22180
(800) 368-3270 (703) 938-3350

Galaxy Electronics
P.O. Box 1202
Akron, OH 44309
(216) 376-2402

Grove Enterprises
P.O. Box 98
Brasstown, NC 28902
(704) 837-9200

Ham Station
P.O. Box 6522
Evansville, IN 47719
(800) 523-7731

ICOM America, Inc.
2380-116th Avenue NE
Bellevue, WA 98004
(206) 454-7619

Radio Shack outlets nationwide

Scanner World USA
10 New Scotland Avenue
Albany, NY 12208
(518) 436-9606

Universal Shortwave
1280 Aida Dr.
Reynoldsburg, OH 43068
(614) 866-4267

Appendix D

Basic Computer Program for Search/Scan Operations

The use of a small, inexpensive personal computer as a search/scan log is very efficient. Although there are database list-management programs available for this application, a degree of memory efficiency and security can be obtained using this simple program:

Section One: Password Access

```
NEW
10 CLS ("HOME" command for Apple)
20 INPUT A$
30 IF A$ = "PASSWORD" THEN LIST 100-180  *
40 IF A$ ≠ "PASSWORD" THEN NEW  **
```

Section Two: REM Statement Storage

```
100 REM FREQUENCY LOG SEARCH/SCAN OPS
101 REM TIME ON:       TIME OFF:   OPERATOR:
102 REM FREQ:          TIME:
```

```
103 REM FREQ:          TIME:
104 REM FREQ:          TIME
```

* You can use any word as your password.
** If the user does not enter the proper password, then the program immediately clears from memory

This format allows maximum memory use, and simple LIST commands can be used to access and add data. You can have several thousand sequentially numbered REM statements for each entry, and you can enter these statements by time or frequency.

Storage on disk or cassette is possible, and the LIST command will provide you with a complete printout of all of your entered programming data.

The operator gets only one chance to enter the proper password. If he fails to key it in properly, the entire system will crash.

Appendix E

Glossary

Abwehr. German intelligence operation during World War II. Abwehr had many major intelligence victories and is still considered one of the best examples of an efficient wartime intelligence operation.

AM. Amplitude Modulation.

APCO. Associated Public Safety Communications Officers. An official nonprofit organization in the United States representing radio and telephone dispatch operators for emergency services.

ATC. Air-traffic control. Communications between aircraft and ground stations regarding flight planning, takeoff and landing instructions, navigation, and flight conditions.

ATM. Automatic Teller Machine. Computerized service window providing bank customers with 24-hour access to banking services.

Baader-Meinhof. West German terrorist organization, also known as the Red Army Faction. Set up by Andreas Baader, Ulrike Meinhof, and others, it was a violent, PLO-trained group responsible for several acts of terrorism in the 1970s. Meinhof hung herself in her jail cell on May 8, 1976, and Baader committed suicide in his cell on October 20, 1977.

Belden 9913. High-grade coaxial cable. Considered one of the best choices for low-loss feedline connections.

Black radio. Term used to define the use of imitative communications deceptions (ICD), when a target believes enemy radio transmissions are originating from friendly sources.

BNC connector. Stable, reliable coaxial cable connector used extensively with radio scanners and hand-held transceivers.

Book code. A simple code using a specific book made available to all parties in a covert network. The code describes the page number and location on the page for each word in the message.

Bootlegging. Criminal use of unauthorized radio equipment and frequencies.

Bug. A small electronic eavesdropping device that transmits audio signals from a target surveillance area to a nearby receiver.

CB. Citizens Band radio, covering the 11-meter HF spectrum from 26.965 MHz to 27.405 MHz.

CIA. Central Intelligence Agency. Coordinator of all U.S. intelligence activities, including defense and foreign policy actions.

Coax. Coaxial cable. Used as interconnection between radio equipment and antenna systems.

CRIS. Computer Radio Interface Systems.

Cutout. A trusted intermediary between two espionage agents or agencies.

CW. Continuous Wave. The interruption of a transmitter's oscillator output with a keyer, which sends a series of tones in the form of "dots" and "dashes" known as Morse Code.

DES. Digital Encryption System. Voice and data scrambling system developed by the NSA for use in radio and telephone traffic. It is a secure, sophisticated system that is difficult to defeat.

DIA. Defense Intelligence Agency. U.S. military intelligence group directed by the Pentagon.

ECCM. Electronic Counter Countermeasures. Methods used to defeat radio jamming operations.

ECM. Electronic Countermeasures. All activities, active and passive, intended to deny the enemy effective use of the radio-frequency spectrum.

ECPA. Electronic Communication Privacy Act of 1986. A U.S. law that makes it illegal to monitor certain types of radio traffic.

Encryption. Using codes and altered terminology to produce text and messages that cannot be read by the enemy.

EW. Electronic Warfare. The use of the radio-frequency spectrum to protect friendly communications while defeating enemy communications.

FBI. Federal Bureau of Investigation. The primary U.S. domestic intelligence agency.

FCC. Federal Communications Commission. The agency responsible for controlling radio and other communications emissions in the United States.

Flash paper. Cloth fiber paper impregnated with incendiary powder that causes the entire paper to "flash" and rapidly disintegrate when exposed to flame or high heat. Available in hobby and magic stores.

FM. Frequency Modulation.

Foil tape. A metallic conducting adhesive tape used in burglar alarm installations.

Frequency counter. Electronic digital receiver that provides the user with an accurate frequency readout of a nearby radio transmission.

F-type connector. Coaxial cable connector used with radio and cable TV systems.

Gisting. Tactical communications term referring to the operator's skill in getting the "gist" or content of an intercepted verbal message, without having to provide a verbatim transcript of traffic.

GRU. Soviet Military Intelligence organization. Chief Intelligence Directorate of the General Staff.

GSG-9. West German counterterrorist organization. Border Protection Unit 9 was set up by the West German government after the 1972 Munich Olympics disaster. GSG-9 is considered to be one of the best counterterrorist units in the world.

Hertz. The unit used to measure frequency, equaling one cycle per second.

HF. High Frequency.

ICD. Imitative Communications Deception. Imitating the enemy's use of radio to deceive and disrupt his activities.

IF. Intermediate Frequency.

IRA. Irish Republican Army.

Jiggling the wire. Term used in surveillance referring to an operation that causes the target to behave in such a manner that the target incriminates himself while being monitored and observed.

kHz. Kilohertz. A unit of frequency equal to 1,000 hertz. Ten kHz is 10,000 hertz.

Line-of-sight. Distance generally considered to be just to the visible horizon. Line-of-sight communications do not refract off the ionosphere except in rare instances.

Listening post. Radio monitoring site set up to listen in on radio communications not intended for the monitor. Also called a monitor station.

MHz. Megahertz. A unit of frequency equal to one million hertz.

MI6. British Secret Service. Military Intelligence Six is similar in mission and function to the CIA.

Mike-keyed. A condition when a microphone is accidently pressed, which ties up the receiver and prevents it from receiving other incoming traffic. Also known as a "units doubling" condition.

Motorola connector plug. The common two-connector plug used for vehicle radio antenna connections and some commercial scanners.

NSA. National Security Agency. The largest and most covert of all U.S. intelligence agencies, the NSA is charged with communications intercept and security worldwide.

N-type connector. Radio plug used for high-quality radio work. Rugged, water- and weatherproof, and very reliable.

NVA. North Vietnamese Army.

OPSEC. Operational Security. All precautions and actions that deny the enemy knowledge of one's intentions or activities.

Oscillator. An electronic device that produces alternating current (AC) power at a frequency determined by certain components in its circuitry. The IF section of a receiver and the typical transmitter are both oscillators.

OSS. Office of Strategic Services. Predecessor of the CIA. An aggressive intelligence and covert operations agency set up by General William "Wild Bill" Donovan during World War II.

Phonetic alphabet. Use of code words to designate each letter in a radio transmission to insure proper understanding of the message.

Pirate. Individual or group that illegally uses portions of the radio spectrum for unauthorized radio transmissions.

PL-259. Commonly used radio coaxial cable connection.

Polarization. The position of a radio antenna, generally either vertical or horizontal.

Propagation. The travel of radio waves through the atmosphere.

Radar. RAdio Detecting And Ranging. System used for determining an object's distance, direction, and speed of travel, using radio waves that are transmitted and then received by the system. The duration of time between transmit and receive is measured and displayed on a screen.

RC. Remote Control.

RCA connector. Standard audio jack used for stereo connections and some consumer radio equipment.

RDF. Radio Direction Finding. Electronic means of locating radio transmissions by using directional antennas and sensitive radio receivers.

RF. Radio Frequency.

RFI. Radio Frequency Interference.

RG-8. Low-loss coaxial cable.

RG-58. Commonly used coaxial cable; much thinner than RG-8.

SAS. Special Air Services. British commando and anti-terrorist group.

SDU. Spectral Display Unit. A feature found on sophisticated intercept and monitor radio equipment that provides video display of a specific portion of the radio spectrum.

SF. Special Forces. U.S. Army unconventional warfare units. The "A-Team" is the basic SF combat unit.

Sky wave. Radio propagation that uses the ionosphere to send radio signals long distances. Also known as "shortwave."

Spark gap. A method of generating RF energy that creates an electrical spark. Seldom used for jamming due to its broad frequency coverage and the use of noise filtering in most receivers.

Spetsnaz. Soviet special forces soldiers. Controlled by the GRU, Spetsnaz troops are assigned the task of covert infiltration, sabotage, radio jamming, and assassination in occupied enemy areas.

TACREP. Tactical Report. A written report that provides information on intercepted radio traffic.

Ten-codes. Brief radio codes beginning with the number 10 that describe specific conditions in law-enforcement communications.

Triangulation. Method of radio direction finding that uses two or more receivers to locate a specific transmitter.

UHF. Ultra High Frequency.

VHF. Very High Frequency.

Voter. Electronic device that takes the strongest signal from a series of satellite repeaters and sends only that signal over the system.

Index